SPECIAL FORCES OPERATIONAL TECHNIQUES

			Paragraph	Page
CHAPTER 1.		GENERAL	1, 2	5
2.		INTELLIGENCE		
Section	I.	General	3–5	6
	II.	Area study/area assessment	6–10	9
	III.	Photography	11–13	16
CHAPTER 3.		PSYCHOLOGICAL OPERATIONS		
Section	I.	Psychological operations and persuasion.	14–17	19
	II.	Planning	18, 19	25, 26
	III.	Propaganda production and use.	20–25	28
CHAPTER 4.		INFILTRATION		
Section	I.	Infiltration in unconventional warfare operations.	26–30	37
	II.	Planning considerations for counterinsurgency operations.	31	46
CHAPTER 5.		PLANNING AND OPERATIONS		
Section	I.	Unconventional warfare	32–42	52
	II.	Considerations for counterinsurgency.	43–48	67
	III.	Training of indigenous forces.	49–52	79

1

		Paragraph	Page
CHAPTER 6.	AIR OPERATIONS		
Section I.	General	53–55	82
II.	DZ selection and reporting	56–60	87
III.	Marking drop zones	61–66	105
IV.	HALO operations	67–71	117
V.	DZ landing operations	72–74	121
VI.	Air landing operations	75–79	128
VII.	Landing zone (water)	80	140
VIII.	Landing zones for rotary-wing aircraft.	81–88	149
CHAPTER 7.	RAIDS AND AMBUSHES		
Section I.	Raids	89	167
II.	Ambushes	90	181
III.	Other interdiction techniques	91	190
CHAPTER 8.	WATER OPERATIONS		
Section I.	General	92, 93	194, 195
II.	Landing operations	94–99	196
III.	Underwater operations	100–103	205
IV.	Small-boat operations in support of counterinsurgency.	104–107	207
CHAPTER 9.	COMMUNICATIONS		
Section I.	Unconventional warfare	108–114	222
II.	Counterinsurgency	115–118	253
CHAPTER 10.	LOGISTICS		
Section I.	External logistics	119–121	260
II.	Internal logistics	122–132	263
CHAPTER 11.	DEMOLITIONS AND ENGINEERING.		
Section I.	Unconventional warfare	133–146	272
II.	Counterinsurgency	147–151	325
III.	Metric calculations	152	333
IV.	Atomic demolition munitions	153, 154	338
CHAPTER 12.	MEDICAL ASPECTS OF SPECIAL FORCES OPERATIONS.		
Section I.	General	155–158	340

		Paragraph	Page
CHAPTER 12.	MEDICAL ASPECTS OF SPECIAL FORCES OPERATIONS—Cont.		
II.	Medical requirements for GWOA's.	159–164	344
III.	Medical requirements for counterinsurgency operations.	165, 166	351
IV.	Veterinary medical techniques.	167–171	355
V.	Collection of intelligence information.	172–174	359
APPENDIX I.	REFERENCES		363
II.	FIELD EXPEDIENT PRINTING METHODS.		369
III.	AIR AND AMPHIBIOUS MESSAGES.		384
IV.	FIELD EXERCISE		389
V.	AREA STUDY		449
VI.	AREA ASSESSMENT		465
VII.	CATALOG SUPPLY SYSTEM.		482
VIII.	EXAMPLES OF MASTER TRAINING PROGRAM FOR INDIGENOUS FORCES.		495
GLOSSARY			512
METRIC CONVERSION TABLES			515
INDEX			519

Special Forces Operational Techniques

ISBN 0-87364-047-0
Printed in the United States of America

Published by Paladin Press, a division of
Paladin Enterprises, Inc., P.O. Box 1307,
Boulder, Colorado 80306, USA.
(303) 443-7250

Direct inquiries and/or orders to the above address.

All rights reserved. Except for use in a review, no
portion of this book may be reproduced in any form
without the express written permission of the publisher.

Neither the author nor the publisher assumes
any responsibility for the use or misuse of
information contained in this book.

CHAPTER 1
GENERAL

1. Purpose and Scope

This manual discusses operational techniques in consonance with doctrine outlined in FM 31-21 which may be used by Special Forces. These methods are applicable to both nuclear and nonnuclear warfare in either unconventional warfare (UW) or counterinsurgency operations. This manual describes intelligence, psychological considerations, infiltration, air operations, amphibious operations, communications, logistics, demolitions, Special Forces field maneuvers, medical aspects and other techniques.

2. User Comments

Users of this manual are encouraged to complement the following chapters with appropriate field manuals and technical manuals listed in appendix I. Users are also encouraged to submit recommended changes or comments to improve this manual. Comments should be keyed to the specific page, paragraph, and line of the text in which changes are recommended. Reasons should be provided for each comment to insure understanding and complete evaluation. Comments should be forwarded direct to the Commandant, United States Army Special Warfare School, Fort Bragg, N. C. 28307.

CHAPTER 2
INTELLIGENCE

Section 1. GENERAL

3. Purpose

This chapter provides information and guidance to Special Forces detachments in the intelligence aspects of operational planning for unconventional warfare and counterinsurgency operations. This chapter will furnish detachment commanders guidelines for analyzing areas of operations through area study and area assessment. Additionally, the operational use of cameras, organic to Special Forces detachments, to improve security and intelligence gathering is included.

4. Intelligence Considerations in a Guerrilla Warfare Operational Area (GWOA)

A thorough knowledge of the enemy, weather, terrain, and resistance potential coupled with an intimate understanding of the population within the operational area is essential to the success of the unconventional warfare mission. Once deployed, the Special Forces detachment is ideally situated to contribute to the overall theater intelligence plan. By using indigenous agencies and sources subordinate to the area command it can assemble and relay to the Special Forces operational base (SFOB) intelligence information of value to the unified and component commanders; however, security of the Special Forces detachment and the resistance efforts restrict radio traffic, thus limiting the amount of intelligence information which can be relayed.

a. The area command is able to exploit sources generally unavailable to other theater forces. The Special Forces commander has available three primary elements to assist in gathering intelligence: the guerrilla force, the auxiliary, and the underground. For detailed information on the functions of these elements see FM's 31-20A and 31-21A.

b. Intelligence systems in the GWOA are primarily geared to support the Special Forces detachment and guerrilla forces in planning and conducting operations, securing bases, and preventing compromise by enemy forces. These systems also assist the auxiliary force in planning support functions and the underground in planning and conducting sabotage and other activities to disrupt enemy activities.

c. One of the basic functions of the area command is the establishment of a sound counterintelligence program. Such a program neutralizes the enemy's intelligence gathering systems and prevents penetration of the guerrilla force by enemy agents. Programs initiated to safeguard and secure the guerrilla's position are—

    (1) Screening of guerrilla personnel and recruits.

(2) Deception operations.
(3) Surveillance of the local populace.
(4) Spreading false rumors and false information concerning guerrilla forces strength, location, organization, training, and equipment.
(5) Determining enemy capabilities and modus operandi.
(6) Penetration of enemy intelligence systems and counterintelligence organizations.

5. Intelligence Considerations in Counterinsurgency

a. In counterinsurgency operations, Special Forces detachments require accurate, detailed, and timely intelligence to successfully accomplish assigned missions. In tactical operations involving the employment of paramilitary forces, particularly in pursuit operations, it is essential that detachments know the present and future location and identification of the insurgent at hamlet, village, district, and province level. Since the nature of guerrilla warfare requires that insurgents have the support of the local population, the detachment commander concentrates his efforts on the population to determine the amount of support it affords the insurgent and its attitude toward the established government. There are several methods in which the Special Forces detachment commander may acquire the information needed.

b. It may not be possible for the Special Forces detachment commander to obtain sufficient information through overt means from the populace. The release of information to the Special Forces detachment by the people could result in retaliation by the insurgent, thereby destroying the usefulness of the people as informants. The detachment commander must consider the use of clandestine intelligence and counterintelligence operational techniques as primary methods for collection. For detailed information on intelligence nets and counterintelligence operations, see FM 31-21A, FM 31-21, FM 31-20A and FM 30-17.

Section II. AREA STUDY/AREA ASSESSMENT

6. Area Study

a. General. Special Forces detachments committed into operational areas to support unconventional warfare and counterinsurgency operations will find, in the majority of cases, that activities in the area are customarily supported and accompanied by extensive political and economic activities. These may be overt or clandestine, conducted by individuals or groups integrated or acting in concert with established resistance forces. In order to improve their chances for success, Special Forces detachments require a greater degree of preparation in predeployment intelligence than normal Army units of battalion or comparable size. To accomplish this, detachments prepare extensive area studies of assigned areas of operations. Although area studies are prepared and provided by special research agencies, the studies prepared by the detachments are organized into a more appropriate format for operational use. There are two area study categories.

b. General Area Study. This is the detailed background knowledge of an area, region, or country. For a sample area study guide, see appendix V.

c. Operational Area Intelligence. This is detailed intelligence of a specific area of operations which includes information acquired from all sources to include—

  (1) Selected personnel exfiltrated from objective area.
  (2) Existing resistance elements.
  (3) Conduct of operations.
  (4) Intensive study of languages and customs.
  (5) Active propaganda programs.
  (6) Study of economic and social aspects of objective areas.
  (7) Conduct of intelligence operations.
  (8) Local law enforcement and intelligence functions (conuterinsurgency only).
  (9) Indigenous forces engaged in combat operations against insurgent forces (counterinsurgency only).
  (10) Military civic action programs (counterinsurgency only).

7. Operational Use in Unconventional Warfare

a. Preparation. The Special Forces group S–2 procures the necessary intelligence documents from which the detachment's general area studies are prepared. Coordination through prescribed channels is effected with all appropriate theater intelligence agencies for the continuous procurement of timely intelligence. Several methods of area study preparation are feasible.

(1) Preparation by operational detachment. This method is more advantageous since the detachment concerned is intimately with its mission and capabilities to accomplish this mission. One major disadvantage is that unit training requirements often limit the time available for detachments to prepare a detailed, comprehensive area study.

(2) Entire preparation by group S-2 on the basis of available information concerning areas of operation.

(3) Preparation by area specialist teams (AST).

(4) Preparation through a combination of these methods with revision by the group S-2 based upon the latest available intelligence.

b. Area Study Preparation Methods. If the detachment prepares its own study from information available to the S-2, the detachment commander has the advantage of assigning specific areas of interest to selected personnel in keeping with their particular specialty. For example, the detachment intelligence sergeant can conduct extensive research and study on the political structure of the area of operations, groups in opposition to the recognized government, and order of battle of the enemy occupying the area. The combat engineer specialist can prepare studies on the major lines of communication and bridges throughout the country by gathering pictures and data on bridge construction, rail yards, major tunnels, and highway nets. This includes information on both primary and secondary roads; surfaces, gradients, and traffic capabilities during all sort of weather. A second method would be for two detachment members of different specialities to research a specific area of interest. An example of this is to have the medical specialist and the combat engineer supervisor research target data while the medical supervisor and combat engineer specialist research area health and medical facilities. This technique provides both cross-training in specialities and also greater dissemination of specific sections before completion of the entire area study. Once these studies have been completed and the detachment personnel responsible for specific sections have prepared their summaries, extensive briefings and orientations are conducted for the remainder of the detachment. These are continued daily until the detachment commander and responsible SFOB staff members are satisfied that each member of the detachment is intimately familiar with the assigned area and its peoples.

c. Operational Area Intelligence. When a detachment is selected for commitment into a specific GWOA, the detachment is placed in isolation in the briefing center of the SFOB and is then ready to receive the operational area intelligence. This is the detailed intelligence of the GWOA from which operational plans are formulated. During these intelligence briefings one source of information referred to as an "asset" may be introduced. The asset is normally a person native to the assigned area of operations who has recently been exfiltrated from that area, thoroughly interrogated for intelligence information, and who volunteered or was recruited to assist in the Special Forces mission. If possible, the asset should have a comprehensive background knowledge of the objective area; it is desirable that he be a member of the resistance movement. When considered necessary he is infiltrated back into the operational area with the detachment to assist on contacting members of the resistance movement.

d. Brief-Back. Once the detachment has completed its preparation for deployment and concluded its studies of the area in relation to its operational missions, the SFOB staff conducts a series of brief-backs. During this period, every member of the detachment is required to brief the group commander, the SFOB staff, and members of the Joint Unconventional Warfare Task Force (JUWTF) on all aspects of their assigned mission until the SFOB commander is satisfied that the detachment is ready for deployment.

8. Operational Use in Counterinsurgency

a. Preparation. Basically the same preparations are made for commitment into a counterinsurgency environment as into a GWOA. However, the Special Forces detachment is required, because of the nature of activities in counterinsurgency operations, to make extensive studies of local customs, religious beliefs, languages, cultural backgrounds, and personalities. Special Forces in counterinsurgency operations need to exploit to the maximum local law enforcement agencies; security and intelligence elements and organizations; and interpreters, guides, and trackers.

b. Study Methods. The detachment itself may desire to prepare the area studies; however, the basic difference in preparing the study for a counterinsurgency situation is that primary emphasis is placed on the study of propaganda appeals, symbols and their uses, and techniques of propaganda dissemination. Gaining attention, understanding, and credibility among the people of the selected area is another primary consideration. Additionally, area handbooks dealing with the sociological, political, economic, and military aspects of the area are researched thoroughly to increase the background knowledge of the detachment before commitment.

c. Operational Area Intelligence. This information normally is given to the detachment upon arrival in the receiving state by the U.S. military advisor's staff and other elements of the Military Assistance Advisory Group (MAAG). Operational intelligence covers specific areas of operation, targets, missions, military operations, civic action, organization and training of paramilitary forces and their employment, border operations, air operations, and other missions designed to solidify host country programs to gain the support of the population in its fight againts insurgency.

9. Area Assessment

a. Area assessment is the collection of specific information by the Special Forces detachment which commences immediately upon entering the area of operations. It is a continuous process which confirms, corrects, refutes, or adds to previous intelligence acquired before commitment. The area assessment is also the basis for changing detachment operational and logistical plans which were made before commitment into the area. Assessment may assume two degrees of urgency—immediate and subsequent. Matters of immediate urgency are included in the "initial" assessment; the "principal" assessment is a continuous collection of information conducted until exfiltration or evacuation from the operational area. An explanation and recommended format for initial and principal area assessments are presented in sections I and II of appendix VI.

b. Major changes in the area study indicated by area assessments will be furnished to the SFOB during the course of normal communications. These changes provide the latest intelligence information to the area specialist teams.

10. Psychological Intelligence Considerations

The area study provides detailed information

concerning the people, religion, customs, and other background information necessary for planning psychological operations in support of Special Forces activities. Additional details on psychological operations intelligence requirements can be found in paragraphs 14 through 17, and in FM 31-21, FM 31-21A, and FM 33-5.

Section III. PHOTOGRAPHY

11. General Uses

The preservation of unit records is one of the more important uses of the detachment camera. These records include such documents as the detachment journal, summaries of operations, intelligence reports, details of enemy atrocities, records of arms and equipment disposition, expenditures of funds, and information concerning indigenous personalities. Photographing these documents and subsequently caching or exfiltrating the negatives provides a method of records preservation and security not obtainable by other means. Special Forces operational detachments will find their organic photographic equipment important in making identification photographs for population control and for organization and control of paramilitary units. To avoid having a large amount of sensitive material on hand, the detachment normally photographs these items at frequent intervals. After processing the negative and determining its acceptability, the originals of unit records may be destroyed.

12. Intelligence Photography

Intelligence gathering activities are facilitated by using the camera, particularly in target reconnaissance. A good negative or print of a target installation gives the detachment an opportunity to make detailed and deliberate studies which often reveal information the casual observer would not have been able to report. This same negative or print also provides a valuable aid in briefing personnel about the installation. It is often reveal information the casual observer to the vicinity of the objective to obtain close-up photographs, and target installations may also be photographed from a distance by using telephoto techniques. Photography provides an excellent means of supplementing reports of captured enemy arms or equipment, because items too large or bulky to evacuate can be photographed in detail and the finished negative exfiltrated as the situation permits.

13. Equipment and Supplies

The camera equipment presently included in the TOE of the operational detachment is adequate for the job intended; however, the following additional accessories are suggested for more satisfactory photographs:

a. A 35-mm developing tank, preferably of the daylight loading kind.

b. Two unbreakable plastic bottles or flasks for chemicals.

c. A small thermometer.

d. A small exposure meter to insure proper settings.

CHAPTER 3
PSYCHOLOGICAL OPERATIONS

Section I. PSYCHOLOGICAL OPERATIONS AND PERSUASION

14. Psychological Operations

One of the critical factors of Special Forces operations in unconventional warfare and in counterinsurgency is the psychological operations effort which supports various Special Forces programs. Basically, psychological operations is concerned with persuading people, or groups of people, to take certain actions favorable to one's interests. In an insurgent situation, psychological operations can be called upon to persuade the people of the area to actively and willingly cooperate with the local government, disrupt the efforts of the insurgent, and assist in separating the insurgent from other elements of the nation's population. In an unconventional warfare role, psychological operations are designed to achieve just the opposite effects: disassociation of the people with the government in power, creation of shared goals for the resistance movement and the population, and mutual help and cooperation between the guerrillas and the people to disrupt the efforts of the common enemy.

15. Persuasion

The fundamental key to a successful persuasion effort is the extent to which the persuader genuinely understands the group being addressed. Because successful persuasion depends on a thorough knowledge of environmental factors which influence the target, as well as a knowledge of what this target group thinks of itself and its environment, it is difficult to specify detailed suggestions because of the diverse areas in which Special Forces may operate. There are general guidelines, however, which will aid Special Forces personnel in a persuasive effort.

a. The first step is to develop an understanding of the situation that currently exists in the area and the reasons why this situation exists. Specifically, what factors present in the area influence the target group. These factors include outside forces and the accepted ways of meeting particular needs such as food or survival. In addition, awareness of the views of the target group regarding these various factors in its environment and the way the group reacts to items associated with these factors is mandatory as are the reasons why the group holds these particular attitudes. Once the group is understood, chances of success are improved.

b. There are a number of other methods which can be used to develop an understanding of the people to be persuaded. The best overall approach is to combine as many of these methods as possible.

(1) *Area studies.* If there are area studies or similar reports available dealing with the area or group of people with whom Special Forces is concerned, these documents will provide a good background on the subject. Many studies may be too general in nature to be accurate with regard to a particular group. Other studies may not be objective and thus present a biased picture of the group.

(2) *Interviews.* In dealings with the group of interests, Special Forces will be able to sound out some of the views that the group holds. This method can provide some indication of group attitudes. The area of interest will be disguised and any suggestion of "correct" answers to questions will be avoided. It is likely that the person being questioned will tend to give answers that he thinks are expected of him.

(3) *Observation.* Careful observation of the daily activities within the group will provide some confirmation of the conclusions reached as a result of analysis of available reports and of interviews with target group representatives. Attention is paid not only to the more obvious activities, but also to some of the more subtle ways members of the group demonstrate their views. Gestures, who is listened to most often and under what circumstances, and the location of items

connected with various aspects of daily life tend to indicate group attitude. In interpreting group activities an open mind is important so that previously conceived conclusions are not confirmed to the exclusion of other truths. The group will not be judged in terms of American values or an interpretation of attitudes made from an American point of view.

(4) *Previously assigned personnel.* It is desirable to discuss the situation with a predecessor, and this discussion can be very beneficial in providing a basic understanding of the strengths and weaknesses of the people as well as persuasive methods that have worked well in the past. Evaluation of previous conclusions, however, will be made to discover errors. Under any circumstance, each Special Forces representative will make an evaluation of the groups with which he deals.

(5) *Validity of conclusions.* Conclusions previously made will be treated as tentative in nature. By doing this a constant reevaluation can be made in the light necessary because some conclusions may be based on tenuous, unconfirmed information.

c. There are a number of approaches to the problem of persuasion. The best one involves changing, if necessary, the predisposition of a target group to react in a particular way to things they sense in their environment—in other words, their attitudes. Having discovered the existing attitudes of the group an examination is made to define characteristics of the attitudes so that changes or modifications can be made.

d. In any given situation there are certain desirable actions for various groups to take. Understanding the overall situation as it applies to any particular group allows better preparation of potentially successful lines of persuasion. These lines of persuasion are based on the environmental factors currently influencing the target group and on current attitudes toward pertinent subjects. These lines of persuasion are then used to influence the group to adopt those desired actions.

16. Guides for Persuasion

Having established the goals of the persuasion effort (actions desired), the next step is to accomplish these goals. How these goals can be accomplished depends greatly on the nature of the situation and the characteristics of the target group. Nevertheless certain guidelines can be established, although they may not apply to all situations. The guidelines are—

a. Use an Indirect Approach. When modifying attitudes, first work on those which are relatively weak and less frequently aroused. In most cases these attitudes will offer less resistance to change than stronger, more frequently aroused attitudes.

Eventually, the stronger attitudes may have to be modified; but by changing the weaker attitudes first, the stronger attitudes may be made weaker.

b. Use a Variety of Approaches. As many lines of persuasion as possible which have a foundation in environmental factors influencing the target and in the attitudes of the target will be used. People possess current attitudes because these attitudes meet current needs. Even within a single target group, individual needs may differ. Consequently, by using as many supporting ideas as possible, there is a better chance of touching on a meaningful line of persuasion for all the members of the target group. The lines of persuasion will be consistent with established policy and the current situation, relate to something of contemporary interest, and be believable to the target group.

c. Use Group Identification. One of the most powerful forces that can be employed is group pressure. The use of group pressure must be carefully thought out to preclude its backfiring. Each situation is analyzed in this regard. The essential feature of this device, however, is that the target group is made aware that other groups, which it respects, favors the advocated action or perhaps that a significant element of the target group itself favors the action. It is important that any such assertion have some real basis so that the target group will believe the assertion.

17. Complexity

In highly complex situations, it is necessary to

persuade several different groups before the real target is ready to take the desired action. In this case, it is necessary to decide which groups are to be persuaded first and by what means. It may be necessary to use certain methods, such as civic action in a contemporary situation, to support verbal lines of persuasion to induce the target to make the desired response.

Section II. PLANNING

18. Intelligence

a. Psychological operations intelligence is essential. Its acquisition and skillful use are a continuing necessity to psychological operations effectiveness. Psychological operations intelligence is concerned with the determination of receptiveness, vulnerabilities, and actual and potential behavior of target audiences before, during, and after psychological operations are directed toward them. It must provide the means to identify and analyze potential audiences, to determine effective message content, to select and employ suitable media and methods, and to assess effectiveness.

b. The requirements for psychological operations intelligence are formulated in detail. A collection plan is prepared and requests are made through appropriate channels immediately after the assignment of missions. Where feasible, information received is processed, employing a psychological operations journal, worksheet, and situation map; and a psychological operations intelligence estimate is prepared.

c. Initial intelligence studies include historical, cultural, and biographical data and material relating to sociological, political, religious, economic, communications, transportation, and military aspects of the operational area. While such studies provide vital and useful background information they rarely provide sufficient detail to permit effective psychological operations in remote areas typical of Special Forces operations. Additional intelligence is necessary and it is often impossible to acquire it until after commitment of the Special Forces detachment into an area.

19. The Target Audience

a. The target audience is the population segment to which the psychological message is directed. Certain propaganda and information efforts are designed to maintain already existing favorable attitudes. Special Forces psychological operations efforts are aimed at producing specific, desirable actions and normally are conducted to overcome attitudes which condition target audiences against taking the desired actions.

b. Understanding the nature of the target audiences and their place in the sequence of psychological operations activities is essential to Special Forces success in remote area operations. The message, the media, and the method employed are built around and derived from the target audience.

c. In remote areas where Special Forces conduct operations, potential target audiences can be of a radically different cultural composition. Extremely small and separate population segments living in isolated villages in the same operational area can possess contradictory customs, different religions, and be dependent upon conflicting economic necessities. Several competing tribes can inhabit a single operational area, and diversity within the same tribe is not uncommon. Although communities are separated by only a few miles, they may have no common interests and no common, cultural orientation with adjoining villages or the major cities of their native countries. Completely different languages may be spoken. Propaganda which might be effective in metropolitan areas can be entirely inappropriate for dissemination in remote areas of the same country.

d. Psychological operations background intelligence in sufficient detail will rarely, if ever, be available before commencement of Special Forces operations in many areas. From the time an area is entered, an urgent requirement exists for psychological operations intelligence to define and analyze target audiences so that meaningful selections can be made. It is likely that the target audiences selected will be based upon information obtained by personal observations and discussions between members of the Special Forces detachment. Environmental conditions affecting potential target audiences and audience attitudes toward these conditions are charted and analyzed.

Based upon these analyses, estimates are made of target audience susceptibilities to psychological operations and of the abilities of target audience members to control and influence others.

e. This last requirement provides the key to many Special Forces psychological operations in operational areas. It is obvious that appeals by modern, mass communications media are ineffective in areas where no radios exist and the literacy rate is low. Population groups may be so organized as to preclude many types of mass appeal, unless a divisive response is sought. Where effective control and influences in villages or tribal units is vested by custom and mutual consent in one or several individuals, such persons may well constitute the one and only potential target audience for psychological operations designed to unify their population groups.

Section III. PROPAGANDA PRODUCTION AND USE

20. The Message

a. The message is the impulse or meaning the sender seeks to pass on to the receiver. While the message is literally received in the sense of being seen, heard, or read by the target audience, this literal reception is no guarantee of its effectiveness. At its terminus, the message competes for the attention of the recipient with numerous other stimuli and events. To produce the desired response, which is the objective of the psychological operations action, each message is created with a distinct purpose in mind and is skillfully designed to accomplish that purpose. In developing the message, assurance is made that it is not based on the social values and experiences of the writer, but on those of the target audience.

b. Messages have substances and form. Themes are the substantive content; the communicative intent of the message. Several different attitude changes are helpful in producing the same desired behavior or action; for example, a defection can be equally promoted by revulsion of the horror of war, distrust of leadership, or homesickness. Each psychological operations message seeks to evoke a specific response conducive to the desired action. A theme simply states the action being sought and the attitude change being promoted to produce the action. An example of a theme is the encouragement of defection by evoking nostalgia.

c. The message can take many forms: words, spoken or written; pictures; objects such as gifts; sounds such as music; movement in the form of pantomime and dance; or a combination of several of these forms.

21. Media

a. Media are the means by which messages are presented. Effective response, as well as literal reception of messages, depends upon their wise choice and employment. This choice is made after careful consideration of the target audience and the theme and form of the message. Mere convenience or availability does not justify and should not influence the use of a particular media. Communication by media of proven effectiveness in the area of operations and to which the target audience is already accustomed and conditioned is likely to be the most effective. But the probable effectiveness of new innovations will not be overlooked.

b. The absence of an elaborate loudspeaker or printing apparatus need not be a handicap to psychological operations in remote areas because face-to-face communications may prove the most effective means, whether other media is available or not.

c. Forms of entertainment which are traditional or popular in the area of operations are excellent potential media. They usually draw full audience attention, are well received, and lend themselves to the communications of psychological operations messages. Such forms as pantomine, dance, and music employ universal sounds and symbols. When Special Forces detachments sponsor such performances the good will and favorable attitudes created can lead to desired actions. Cautions will be exercised to avoid having the audience identify with the Special Forces detachment and not the host government.

d. Gifts can be used to carry propaganda messages. Such items as soap, matches, salt, needles and thread, seeds, clothing, and other items of utilitarian value make suitable gifts which are acceptable for general distribution. These gifts should be printed on or be wrapped in a piece of paper containing a propaganda message or a symbol which conveys the meaning desired. In selecting gifts, be sure that the gifts are useful and that the use of symbols or messages upon them do not antagonize the receiver. For example, it may be in poor taste to have a piece of soap wrapped in a leaflet containing a copy of the recipient's national flag. The sender must be aware of countermeasures which the enemy may take, such as giving gifts of food which is poisonous and attributing the gifts to the friendly forces.

22. Media and Target Audience

It should be recognized that employment of sophisticated media is backward areas can be ineffective where target audiences are unaccustomed to their use, and that clear and intelligible messages can be misunderstood. Nevertheless, the mere possession and public use of modern communication devices can raise the prestige of the user in the eyes of remote area target audiences. In the use of any media, it is imperative that it be considered carefully so that it does not create an unintended effect on the target audience. There are certain rules for face-to-face communication which have been developed by experience.

a. Avoid dogmatism at all times. The ideas of others are respected in successful communication.

b. Stress accord and approval. A sincere approval of a people's values develops a strong basis for further communications.

c. Avoid minor disagreements if possible.

People like to win discussions, and it is often better to overlook minor disagreement in order to develop stronger rapport and a basis for friendship.

d. Follow and use the audience's line of reasoning at all times. Special Forces personnel, working with people of other cultures, must understand different values and experiences involved. People's concepts of time or right and wrong may differ from the outlook of Americans. They may not understand why U.S. personnel continue trying to accomplish a task against all odds.

e. Be reasonable.

f. Use symbols and language patterns understood by the target audience that will produce the desired behavior attitudes.

23. Assessment of Effectiveness

a. The effectiveness of psychological operations are necessary to determine results. In Special Forces operations, partial or complete estimates of psychological operations effectiveness can be based upon conferences, conversations, and personal observations conducted by Special Forces personnel themselves.

b. Where psychological operations missions require a series of actions over a long of time, the audience behavior desired normally will be extremely difficult to discern and estimate. In these instances, psychological operations intelligence requirements for assessing effectiveness are extensive and continuing. Assessments provide the basis for adjusting and improving methods, developing and revising plans, and setting new psychological operations objectives.

24. Propaganda Development

Themes (lines of persuasion) and symbols selected for use in propaganda are based upon results of target analysis. Themes will be consistent with policy, existing conditions, and other actions of the sponsor. Themes should be directed towards the underlying attitudes of the selected target rather than toward its overt behavior. Both themes and symbols will be meaningful to the target in terms of its view of reality rather than in terms of what the propagandist views as truth and reality.

25. Propaganda Production and Dissemination

The form and content of the actual propaganda product will depend largely upon the considerations discussed in the preceding paragraph. The communications media selected to carry these themes and symbols will depend again upon the target's frame of reference or field of experience as well as upon the resources and capabilities of the propagandists. The operational detachment commander will, in most cases, be forced to resort to his imagination and ingenuity in determining the most meaningful and effective means of communicating the propaganda message to the target audience. Some means and media that can be considered are—

a. Counterinsurgency Operations.

(1) Graphic and visual materials. All forms of printed materials such as newspapers, magazines, pamphlets, posters, and leaflets; motion pictures and photographic slides.

(2) Loudspeakers and public address systems.

(3) Radio and television.

(4) Stage productions, rallies, and other forms of large gatherings.

(5) Smaller, formal and informal gatherings and meetings.

(6) In some situations, the considerations given in *b* below, may also apply.

b. Unconventional Warfare. Access to sophisticated media within the GWOA normally will be limited.

(1) *Graphic and visual materials.* These can be printed materials reproduced on mimeograph and similar lightweight reproduction equipment such as, jelly rolls and other field expedient, reproduction equipment (see app II) or by underground press facilities. Hand produced materials such as letters, posters, wall signs, markings, and symbolic devices such as displays or physical mockups of resistance symbols are used.

(2) *Loudspeakers and other forms of public address systems.* These may be brought

into the GWOA from outside sources, locally procured, or fabricated. Bulk power requirements and security frequently restrict the use of this means.

(3) *Radio, both clandestine and from outside the GWOA.* The technical problems of establishing and operating a clandestine radio, as well as security, restricts the use of this means. If the target can receive radio broadcasts, the use of a clandestine radio broadcast system can be highly effective.

(4) *Inter-personal communications.* In view of the limited public communications media accessible within the GWOA, much of the communication of propaganda messages is through inter-personal exchange. In effect, all members of the Special Forces operational detachment and the resistance forces are used as active propagandists within the limits of security. Depending upon the situation and the target audience, inter-personal communications can be the most effective means of communicating propaganda messages.

(5) *Psychological operations.* Though the task is difficult, psychological operations communication is accomplished among semi-literate and illiterate target groups in varying degrees. The target audience interprets the message in terms

of previous experience and learned response; therefore, it is necessary to understand the experience of the target audience before the intended meaning is communicated. Lack of experience with Western forms of communication is overcome by techniques of presentation that fit their experiences. As an example, the photograph does not communicate meaning to many groups and yet through other approaches communication is accomplished. Drawing on the ground, cut-outs of paper and other materials, and scale models may assist in overcoming an inability to understand a photograph. The key to communicating at this level of literacy is the technique of presentation which uses the symbols, language, and experience of the target audience to express an idea in a way that they understand.

CHAPTER 4
INFILTRATION

Section I. INFILTRATION IN UNCONVENTIONAL WARFARE OPERATIONS

26. General

The success of Special Forces air and amphibious operations in support of unconventional warfare and counterinsurgency operations is primarily dependent upon detailed planning and preparation. This chapter is concerned with the planning preparation, and techniques employed for infiltration into Guerrilla Warfare Operational Areas (GWOA) (para 27–30) and counterinsurgency and counterguerrilla operational areas (para 31). Techniques to be discussed are air infiltration, including high altitude low opening (HALO) techniques and the "blind drop" technique; water infiltration, land infiltration; stay-behind operations. The techniques involved in training indigenous personnel are generally the same as those for any other personnel and are outlined in detail in TM 57–220.

27. Air

a. Air delivery by parachute is one of the principal means available for the infiltration of

Special Forces detachments. In preparing a detachment for infiltration by parachute, consider these factors—

    (1) *Aircraft capabilities.* The maximum number of personnel and amount of equipment that can be delivered, together with the method of dropping, depend upon the capabilities and limitations of the particular aircraft used. Dimensions and other data on U.S. Army and USAF aircraft are presented in FM 101–10, part I, and in FM 31–73.

    (2) *Reception committee.* The presence of a reception committee on the drop zone (DZ) influences the amount of accompanying equipment and supplies as well as the initial actions of the detachment. When a reception committee is available, sterilization of the DZ and disposal of parachute equipment is a lesser problem than when a blind infiltration is conducted. When a reception committee is present, additional equipment and supplies, beyond immediate requirements, may be dropped with the detachment.

    (3) *Equipment and supplies.* The detachment must have in its possession the equipment with which to accomplish initial tasks. These items normally consist of radios, individual arms, and operational TOE equipment which may include medical kits, photographic equipment, binoculars, compasses, TA cloth-

ing and equipment in keeping with climatic conditions in the operational area, and food and survival equipment. The equipment and supplies to accompany the detachment may be dropped using one of the following techniques:

(*a*) *Air delivery containers.* All detachment equipment and supplies are rigged in air delivery containers. They may be dropped as door bundles or by some mechanical means. This technique permits the individual parachutist to jump unencumbered by excess equipment; however, it may result in the loss of valuable items of equipment if the containers are not recovered. This technique should be used only when an adequate reception committee is assured, or in low level drops (500–700 ft) where dispersion is less of a problem and there is little time to release a rucksack.

(*b*) *Air delivery containers/individual loads.* Essential items such as radios are "jumped" on detachment members while less important items are rigged in air delivery containers and dropped as outlined in (*a*) above. Detailed information on containers sizes and aircraft dimensions are listed in TM 57–210 and in FM 101–10.

(*c*) *Individual loads.* Detachment equipment and supplies may be "jumped" as individual loads. This restricts the amount that can be dropped but precludes the loss of essential items through failure to recover containers. This technique is best suited for blind infiltration or when the availability of an adequate reception committee is doubtful. The present method of dropping individual loads consists of packing all items in a rucksack to be released and suspended a safe distance below the jumper; the rucksack landing before the jumper.

(4) *Control.* The detachment commander places himself in the optimum position in the stick to control his detachment. Rehearsals, if necessary, should be conducted to insure the detachment's proper assembly on or off the DZ. Team recognition signals must be clearly understood. Such signals should not be confused with those prearranged with the reception party.

(5) *Ground assembly.* Each member of the detachment is thoroughly briefed on assembly procedures. This includes actions of the individual when approached by guerrillas, i.e., exchange of recognition signals, the location of an assembly point to be used, and the location of primary and alternate assembly points should individual jumpers fail to make contact with the reception committee.

The primary assembly point should be referenced to an easily recognized terrain feature and provide sufficient concealment to allow individuals to remain undetected until such time as they can be recovered. It should be located 100 to 200 meters off the drop zone. An alternate assembly point must satisfy the same criteria as the primary as regards recognition and concealment; but in addition, it should be located 5 to 10 kilometers from the DZ. In addition, each detachment member is carefully instructed concerning disposal of individual parachute equipment and the techniques of erasing signs of the drop.

(6) *Emergency plans.* Consideration is given to the possibility of inflight emergencies, particularly in deep penetration flights. The detachment receives a preflight briefing on the route to be flown and is informed periodically of flight progress. Before enplaning, simple ground assembly plans for such contingencies are established. Should such an emergency arise, the detachment commander, considering the instructions contained in his operation plan and the relative distances to both the infiltration DZ and friendly territory, decides either to continue to the original destination or attempt exfiltration. An emergency plan

should also be provided for use in case of enemy contact on the drop zone.

b. HALO Operations. When enemy air defense discourages normal infiltration by air, parachute entry from very high altitudes may be necessary. Whenever this type of operation is planned in denied areas protected by enemy radar and other detection devices, a system of jamming or disrupting these systems should be established. An important consideration is the availability of aircrews trained in working under arduous conditions in depressurized aircraft at jump altitudes in excess of 33,000 feet. Once HALO parachutists have exited the aircraft, a system for freefall assembly in the air before opening the parachutes must be devised. This is particularly important at night or when conditions preclude visual contact with DZ markings. Assembly aids include special marking devices and materials, visible at night, applied to pack trays, backpacks, and other designated equipment. Other operational characteristics of the HALO technique are presented in paragraphs 67 through 71.

c. Blind Drop.

(1) Selected U.S. and indigenous personnel (commonly called assets) may be air dropped during the initial infiltration phase on drop zones devoid of reception personnel. This technique is referred to as a "blind drop," and may be employed when a resistance element of sufficient size and nature to warrant exploitation

is known to be in the area. In all probability, the force will be small, passive in nature, untrained, but receptive to outside support. Other interested government agencies were either unable or did not have the time and means to train the resistance element in DZ operations. Additionally, the enemy situation might preclude normal DZ markings and recognition signals.

(2) Once the DZ is selected within the operational area, the Air Force has responsibility for flight planning, IP selection, and crew procedures throughout the flight. Normally the drop will be made on a computed air release point or on a visible, selected impact point. If HALO techniques are employed and the ground is not visible, the high altitude freefall release system is used. The dropping personnel employ "tracking" procedures to glide into and select their impact point. (TM 57–220).

(3) Once on the ground, personnel move to a selected assembly area and establish security. The unit commander, along with his asset, then attempts to make contact with the local resistance elements. On the basis of the detachment commander's assessment of the area after contact has been made, he is then in a position to recommend to the SFOB and the JUWTF the feasibility of or-

ganizing the area and committing additional Special Forces units.

28. Water

a. Water offers another practical means for infiltration into operational areas having exposed coastlines. Considerations for water infiltration include—

(1) *Craft limitations.* The characteristics and limitations of the craft largely determine the landing techniques (ch 8).

(2) *Reception committee.* The presence of a reception committee influences the actions of the detachment after landing and the amount of equipment and supplies that may be taken.

(3) *Equipment and supplies.* Adequate waterproofing should be provided to protect supplies and equipment from the effects of salt water. If no reception committee is expected, the amount of equipment and supplies to accompany the detachment is restricted to those quantities that the detachment can transport unassisted. When fleet-type submarines are used, all items are packaged in size and configuration to be passed through the narrow access openings (64 centimeters in diameter) into the pressure hull.

(4) *Ship-to-shore movement.* Assignment of

boat teams, distribution of equipment and supplies, methods of debarkation, and means of navigation to the landing beach are carefully planned. In addition, consideration is given to methods of recognizing the reception committee and disposing of the landing craft.

b. Water infiltration operations normally terminate in a land movement phase.

c. Infiltration by means of a sea plane landing on large lakes, rivers, or coastal waters may be possible. In such a case, infiltration planning by the detachment considers the ship-to-shore and subsequent land movement characteristics of water infiltration operations.

29. Land

Land infiltration is conducted similar to that of a long-range patrol into enemy territory. Generally, guides are required. If guides are not available, the detachment must have detailed intelligence of the route, particularly if borders are to be crossed. Routes are selected to take maximum advantage of cover and concealment and to avoid enemy outposts, patrols, and installations. The location and means of contacting selected individuals who will furnish assistance are provided to the detachment. These individuals may be used as local guides and sources of information, food, and shelter. Equipment and supplies to be carried will necessarily be restricted to individual arms and equipment and communications equipment.

30. Stay-Behind

Special Forces detachments may be preplaced in proposed operational areas before these areas are occupied by the enemy, providing the opportunity to organize the nucleus of a guerrilla force. Stringent precautions are taken to preserve security, particularly that of the refuge areas or other safe sites to be used during the initial period of occupation. Information concerning locations and identities within the organization are kept on a need-to-know basis. Contacts between various elements use clandestine communications. Dispersed caches, to include radio equipment, are pre-positioned when possible. Due to the inadvisability of Special Forces detachment members to function as intelligence agents in urban areas, stay-behind operations normally have a better chance of success in rural areas. When stay-behind operations are attempted in areas of heavy population the detachment will be completely dependent upon the indigenous organization for security, the contacts required for expansion, and the buildup of the effort.

Section II. PLANNING CONSIDERATIONS FOR COUNTERINSURGENCY OPERATIONS

31. General

Infiltration techniques employed in counterinsurgency operations will depend upon assigned missions, number of personnel committed, and availability of suitable transportation. Consideration will be given to air infiltration and will include rappelling from helicopters, HALO operations, and the employment of both fixed-wing and rotary-wing aircraft of the aviation company, Special Forces group water infiltration in areas contiguous to coastal areas, land infiltration, including long-range patrol actions; and stay-behind operations.

a. Air. Air delivery of equipment and personnel by parachute use the same techniques as those used in unconventional warfare operations. Other operational techniques may be employed effectively because counterinsurgency operations include short-range penetrations to objective areas; lack of sophisticated enemy air defenses; and penetrations in remote, inaccessible, insurgent controlled areas.

  (1) In areas inaccessible to normal entry, rappelling from a helicopter can provide access in many cases and enhance the commander's freedom of action in assigned tactical missions in support of counterinsurgency operations. Examples—

   (*a*) Conducting raids against enemy camps and strongpoints.

   (*b*) Establishing blocking positions at designated points while conducting encirclement operations.

   (*c*) Augmenting strike force units in pursuit of insurgent forces.

   (*d*) Infiltrating selected key personnel,

e.g., medical specialist, forward air controller.

(e) Assisting in distressed areas where normal approaches are denied.

(f) Conducting military civic action in very remote isolated areas.

(2) The number of personnel to be infiltrated into an area using helicopter rappelling techniques is limited by—

(a) Allowance cargo of the helicopter.

(b) Hovering ability of aircraft.

(c) Wind conditions and other weather factors.

(3) HALO infiltration may be desirable when a limited impact area is available. The HALO parachutists, using the techniques of "tracking" combined with the maneuverability of the parachute, selects and lands in relatively small areas. HALO operations may be successfully used in deep penetrations for reconnaissance and intelligence missions, locating and firing enemy redoubts and sanctuaries, and locating and establishing suitable DZ's and LZ's for receiving larger attack forces.

(4) Army aviation supporting Special Forces infiltration and other operations may employ airplanes or helicopters to conduct aerial delivery of personnel and equipment by parachute, to conduct air landings, parachute resupply, and low

level extractions. Infiltration of Special Forces elements into insurgent-controlled areas may also be accomplished by low flying helicopters using various ruses and landing techniques which confuse the enemy as to the true location of the drop. This technique is employed when dropping selected, intelligence reconnaissance teams in designated areas to locate secret bases, fix locations of supply depots, and to locate and destroy communications centers or other key installations (ch 5 and 6).

b. Water. Water infiltration techniques used by Special Forces units in support of counterinsurgency operations are the same as those employed in unconventional warfare operations (ch. 8). Since Special Forces operations in support of counterinsurgency will involve limited unconventional warfare operations, with only limited contact and support by local resistance forces, certain operational techniques will vary. A basic difference will be in the lack of any reception committee on the beaches to receive the Special Forces detachment. Therefore, selected naval units will reconnoiter, select, and mark landing sites and direct the loading and unloading of infiltration personnel. The lack of sophisticated beach defenses and radar facilities will permit greater freedom of operations by naval support craft and personnel. A wider variety of missions, not normally associated with Special Forces detachments, may be assigned and carried out with support by other naval forces. These missions may include—

    (1) Psychological activities against selected targets.

    (2) Operations to destroy enemy forces and facilities in conjunction with paramilitary forces.

    (3) Reconnaissance to locate bases and supply depots.

    (4) Gathering intelligence to locate and fix insurgent forces.

    (5) Water operations involving the employment of small boats on rivers, inland waterways, estuaries, and lakes (see ch 8).

c. Land. Land infiltration techniques will be the same as those employed by Special Forces in unconventional warfare operations and long-range patrol actions. Depending upon mission requirements, the Special Forces detachments will be deployed with larger, more heavily armed units such as strike forces, to infiltrate insurgent controlled areas. This technique will differ from a regular unconventional warfare infiltration mission in that the detachment will—

    (1) Not be deployed as a complete detachment.

    (2) Attempt deep penetrations for raids and operations against selected targets.

    (3) Conduct behind the lines operations against targets of opportunity for designated periods of time.

(4) Capture and hold key terrain for the establishment of blocking positions in support of an overall counterguerrilla operations.

(5) Conduct deep penetrations into denied areas and return to friendly areas after dropping selected intelligence and reconnaissance teams for stay-behind operations.

d. Stay-Behind Operations. Special Forces detachments employed in support of counterinsurgency operations and psychological operations in remote areas, working with minority and tribal groups, have a distinct advantage and opportunity to establish and prepare guerrilla operational areas. Detachments employed to organize and train paramilitary units, such as civil defense forces, can prepare this force as the nucleus of a cadre for conversion to a guerrilla organization in the event the enemy overruns and controls the area. The detachment, through prior planning will locate likely DZ and LZ sites, organize and train selected personnel for auxiliary and underground functions, establish supply caches, establish communications facilities, and establish safe areas.

CHAPTER 5
PLANNING AND OPERATIONS

Section I. UNCONVENTIONAL WARFARE

32. Planning for Guerrilla Warfare

Organizational concepts, resistance force relations, psychological operations, intelligence production, logistical support administration, and establishment of training programs in support of guerrilla warfare are explained in chapter 8, FM 31–21. Operational techniques in planning, organizing, training, and conducting operations in support of guerrilla warfare are discussed below.

33. Command and Control

a. Unconventional Environment. Operational detachments are selected for commitment by the Special Forces group commander on the basis of mission assignment and operational readiness. Once committed into the GWOA, operational detachments may have the use of one channel of communication for both operational direction and logistical support or, if deemed advisable and in accordance with local SOP's, have two channels of communications: one for logistical support and one for operational direction. Several methods may be used for control and support of detachments committed into GWOA's:

(1) Operational detachments committed individually are directly responsible to the SFOB for all operations. The detachment contacts the logistical support element direct to request necessary supplies and materials.

(2) When two or more detachments are infiltrated into the operational area and one detachment is a command and control (B or C) detachment with the responsibility of establishing the area command, then operational control is normally established in one of two ways:

  (*a*) Both detachments are in direct contact with the SFOB for operational control; or both detachments have separate supply channels.

  (*b*) The command and control (B or C) detachment exercises operational control over the subordinate detachments, receiving orders from the SFOB and relaying them to the subordinate units. In this case, however, detachments, for security and rapid action, will maintain individual logistical channels for supply requests.

(3) Because of the operations and the distances involved, control measures are not as effective within an area command as they are in a conventional military

organization. Certain criteria are established to increase effective control.

(a) *Operation order.* Sufficient guidance to subordinate units is outlined in the operation order to cover extended periods of time. This is especially true when operations preclude frequent and regular radio contact. Operation orders for Special Forces will include long-term guidance on such matters as psychological operations, (time phasing of interdiction operations), intelligence, air support, external logistical support, evasion and escape, and political and military relationships with the resistance.

(b) *Standing operating procedures (SOP).* SOP's standardize recurring procedures and allow the detachment and the SFOB to anticipate prescribed actions when communications have been interrupted, such as putting into effect emergency resupply operations.

b. Command and Control. Command and control of operational detachments and established guerrilla units in operational areas during and after link-up operations with conventional forces are covered in paragraphs 95 and 96, FM 31-21.

34. Selection of Operational Detachments

When the GWOA has been designated by the

unified commander, any of the Special Forces operational detachments (A, B, or C) may be selected to infiltrate first. One or more may be chosen, and some of the factors influencing the selection of the type and number required are—

a. Character of the Resistance Movement Within the Area. The size and composition of the resistance movement may not be known; or it may be known to be extremely small and unorganized but with a potential for expansion under proper guidance and logistical support. In either event, the immediate infiltration of an A detachment to begin the initial organization and development of the resistance movement may be in order. In another situation, the resistance movement may be highly organized and, except for logistical support and coordination of guerrilla force activities, will require little additional assistance from Special Forces units. Situations may develop where the known leader of the resistance movement is of such importance or caliber that a senior Special Forces officer and a more complex staff will be required to effect the necessary coordination and future development of the force. At this time a B or C detachment may be chosen for infiltration.

b. Existing Situations. The terrain, the enemy situation, complex political problems, or the ethnic groupings within the resistance movement may require two or more detachments to be infiltrated simultaneously. If the GWOA is relatively large and compartmented for security, it is preferable to have several detachments placed into the area initially in order to organize, develop, train, equip, and direct the efforts of the guerrilla force. Regardless of the number of detachments initially infiltrated into a specified area, infiltration of additional operational detachments may be necessary because of increased operations, expansion of the existing guerrilla force, political reasons, or special operations against selected targets.

35. Preparation for Deployment

a. The detachment's thorough, extensive area studies of the operational area provide political, social, economic, and military knowledge of the area and an understanding of the ethnic grouping, customs, taboos, religions, and other local mores. These considerations affect the unit organization, command and control, area of operation, discipline, and the selection of leaders within the guerrilla force.

b. Although maximum use of improvisation must be made, in all phases of operations, the following items accompanying deployed detachments may prove useful in conducting rudimentary training as discussed later in this chapter:

    (1) Grease pencils and colored chalk.

    (2) Target cloth (blackboard substitute).

    (3) Basic manuals on weapons generally found in the area.

(4) Lesson plans for such basic subjects as field sanitation, first aid, map reading, marking of LZ's and DZ's, establishment of security zones, camouflage, dispersion, and simple communication.

(5) GTA's improvised from parachute silk or other such material.

(6) Other similar items of particular value in training the indigenous force. See appendix VII and ATP 31–105.

36. The Guerrilla Warfare Operational Area (GWOA)

A well organized GWOA insures close coordination between operational detachments and resistance elements. After infiltration, the major task facing the operational detachments will be to develop all resistance elements into an effective force. There are several techniques which will facilitate this development. They are completed concurrently as the organization and development of the area progress.

a. General. Establishment of a working and command relationship between the Special Forces detachments and the resistance elements in the area is the initial requirement. A sound working and command relationship helps to develop a high degree of cooperation and some degree of control over the resistance. Control over the resistance force is insured when resistance leaders are receptive to orders and requirements necessary to accomplish the theater mission.

b. Establishment of Security, Intelligence, and

Communications Systems. The detachment stresses that proper organization and development can only be assured through a strong security system. Effective intelligence systems, initially small, must provide timely intelligence in order for the resistance force to react when under pressure from the enemy. Communications systems will be small and quite unsophisticated, but they must be effective and secure. See FM 31-21, FM 31-21A, and FM 31-20A.

c. Establishment of Administrative Systems. Administrative systems should be simple and effective and established early in the stages of development. Administrative systems should include, as a minimum—(1) supply accountability (serial numbered items); (2) personnel rosters; (3) registers of sick, wounded, and deceased; (4) awards and decorations; and (5) a daily journal. Records that can compromise the detachment if they fall into enemy hands may require photographing and caching for safe keeping. Written operations orders and reports will be kept to a minimum and issued on a need-to-know basis.

d. Establishment of Training Programs and Facilities. Training will require a maximum and continuous effort on the part of the detachment. The level of resistance force training must be determined, and training programs must be designed to provide and improve common levels of training. Training programs should be simple but effective with training areas secure from enemy observation and action.

e. Plan and Execute Combat Operations. The selection, planning, and execution of combat action should insure maximum success with a minimum of casualties to the guerrillas. Combat operations should be commensurate with the status of training and equipment available to the resistance force. As training is completed and units are organized, more complex and larger operations are planned and executed.

f. Expand the Resistance Force. Initially, the force may be quite small. The area command constantly analyzes and reviews all previous objectives and develops the area as the resistance force capabilities improve and expand. The expansion program is enhanced by sound logistical systems and successful training programs.

g. Establish Logistical Support Systems. Logistical support systems are categorized as internal and internal logistics. The detachment commander is responsible for an effective, internal supply system which will encompass organization, acquisition, control of supplies, battlefield recovery systems, bartering, emergency caches, and accountability procedures. External supplies are requested from the SFOB by the detachment. These supplies are provided by the sponsoring power and the detachment commander should request only those items that are necessary. The judicious use and control of these supplies is a means of exerting some control over the resistance elements to gain their cooperation and support U.S. objectives.

37. Area Organization

The command structure and the physical organization of the area is a priority task of the Special Forces commander. In some situations the organization of the area may be well established; but in others, organization is lacking or is incomplete. In all cases some improvement in physical dispositions is probably necessary. Organization of the GWOA is dictated by a number of requirements, but it depends more upon local conditions than upon any fixed set of rules. Factors to be considered are effectiveness of guerrilla organization, extent of cooperation between resistance forces and local civilians, enemy activity, and topography.

a. Area Complex. After his initial assessment, the detachment commander may organize his operational elements into an area complex to achieve dispersion and control.

(1) *Definition.* An area complex consists of guerrillas bases and various supporting elements and facilities. Normally included in the area complex are security and intelligence systems, communications systems, mission support sites (MSS), reception sites, supply installations, training areas, DZ's, and LZ's.

(2) *Characteristics.* The complex is not a continuous pattern of tangible installations but a series of intangible lines of communications emanating from guerrilla bases connecting all other resistance

elements. The main guerrilla base is the hub of a spiderweb-like complex which is never static but constantly changing.

b. The Guerrilla Base. The basic establishment within the GWOA is the guerrilla base.

 (1) *Definition.* A guerrilla base is a temporary site where headquarters, installations, and units are located. There is usually more than one guerrilla base within an area complex.

 (2) *Characteristics.* From one base, lines of communication connect other bases and various elements of the area complex. Installations normally found at a guerrilla base are command posts, training areas, supply caches, communications, and medical facilities.

c. Locations. By virtue of their knowledge of terrain, guerrillas can recommend the best areas for locating various installations. Remote or inaccessible areas are ideal for the physical location of guerrilla camps; however, the lack of these remote areas does not prevent guerrilla operations. Approaches to the base, well guarded and concealed, are revealed only on a need-to-know basis. Alternate base areas must always be established for mobility and flexibility dictate the location of guerrilla installations.

38. Civilian Support

For complete details on the organization, missions, capabilities, and role of civilian support within the GWOA, refer to FM 31–21, FM 31–21A, FM 31–20A, and FM 41–10.

39. Intelligence and Security

For complete details on intelligence requirements within the GWOA and the security required to conduct successful operations, refer to FM 31–21, FM 31–21A, and chapter 2 of this manual.

40. Operations

a. Major emphasis is placed upon interdiction operations—denying use of selected areas to the enemy, and destroying facilities, military installations, and equipment. Interdiction may be the destruction of one vehicle by one individual, or attacks by larger groups or forces against strategic industrial areas or sites. When properly coordinated and conducted with other activities in the enemy's rear areas, interdiction operations can make significant contributions to the destruction of enemy combat power and his will to fight. Although tactical in execution, interdiction operations have a strategic objective and have both long-range and immediate effect upon the enemy, his military force, and ultimate population support. Ultimate goals of directed guerrilla actions are to—

    (1) Destroy or damage vital installations, equipment, and supplies.

    (2) Capture supplies, equipment, and key enemy personnel.

under no circumstance does the guerrilla force allow itself to become so engaged that it loses its freedom of action and allows enemy forces to encircle and destroy it.

g. When faced with an enemy offensive of overwhelming strength, the commander may disperse his force, either in small units or as individuals, to avoid destruction. This course of action should not be taken unless absolutely necessary for it renders the guerrilla organization ineffective for an undetermined period of time.

Section II. CONSIDERATIONS FOR COUNTERINSURGENCY

43. General

In counterinsurgency operations, U.S. supported forces operate in less restrictive environments than in unconventional warfare, and their efforts are directed towards countering insurgent movements by denying them the support of the population and by destroying them by combat actions.

44. Missions

Missions assigned TOE Special Forces detachments committed into counterinsurgency operational areas are broad in scope. Once in-country, these missions are further broken down into detailed requirements dictated by the local situation and the counterinsurgency plan for that area. Missions may be to—

a. Train, advise, and provide operational assistance to indigenous Special Forces detachments, ranger-type units, paramilitary forces, and other military forces.

b. Perform limited, military civic action and environmental improvement programs to support the overall counterinsurgency plan. For additional information on civic action see FM 31-73 and FM 41-10.

c. Organize, train, advise, and direct tribal, village, and other remote area groups in counterinsurgency operations. This may include establishment of external defenses and internal security, border operations, and surveillance tasks.

d. For details on additional missions assigned Special Forces detachments supporting counterinsurgency operations, refer to FM 31-21, FM 31-21A, FM 31-20A, FM 31-73, and FM 41-10.

45. Application of UW Techniques

Unconventional warfare techniques, in establishing intelligence nets; evasion and escape mechanisms on a limited scale; the use of psychological methods to gain support of the local population; and raids, ambushes, and air operations all have application in counterinsurgency. Techniques employed will depend largely upon the assigned tasks and the support required. The organization and presence of effective local defense units can neutralize the insurgents' efforts to gain support from the people. Special Forces detachments must carefully analyze each mission assigned and evaluate them in the light of unconventional warfare techniques and their application to counterinsurgency operations. For additional information refer to FM 31-21, FM 31-21A, and FM 31-20A.

a. Civil Guards. Special Forces detachments will find that civil guard units are primarily charged with the mission of internal security. The civil guard is normally trained in individual weapons, light machine guns, and small mortars. The civil guard, in its security role, performs limited tactical missions such as raids, ambushes, and the pursuit of insurgent forces. The civil guard usually is organized into companies and battalion-size units. The tactics and techniques that Special Forces units use in support of guerrilla operations have the same application as when training civil guards. The Special Forces detachment commander can receive support in counterinsurgency not normally available in guerrilla operations. This support includes artillery fire support; armed light aviation support; close air support; extensive communications; and effective medical evaluation.

b. Self-Defense Units. These units normally are responsible for the security of villages and hamlets; guarding major headquarters, bridges key intersections, and local airstrips; and for conducting limited, offensive operations. They may be organized into platoons or squads, and members normally are from the villages and hamlets within the area. With proper training they can conduct around the clock patrols, raids, and ambushes. In nonmilitary missions they can assist in emergency relief and be the principle support of environmental improvement, self-help programs set up by Special Forces units.

c. Civil Defense Groups. These groups are more likely to be identified with primitive tribes in remote areas, not readily accessible to regular forces. Among those included in this group are people from rural areas, ethnic minorities, and other miscellaneous groups such as workmen's militia, youth organizations, and female auxiliaries. They can provide local and internal security of their villages and hamlets when properly trained and armed with adequate weapons. Training emphasis is on defensive tactics. Special Forces detachments assigned to these groups, especially in remote and border areas, will conduct extensive training in guerrilla operations.

This includes training in—
  (1) Hunter-killer team techniques.
  (2) Trail watching.
  (3) Border surveillance.
  (4) Ambush of supply routes.
  (5) Raids on insurgent camps.
  (6) Intelligence gathering penetrations of insurgent controlled areas.

46. Selection of Operational Detachments

The same general criteria applies in selecting detachments for counterinsurgency operations as for unconventional warfare operations. The same preparations are made for infiltrating counterinsurgency areas except that logistical support is more rapid and secure, lessening the amount of equipment accompanying the detachment. For general guidance on the training and advisory capabilities of Special Forces detachments, see chapter 11, FM 31–21.

47. Operations

a. Major emphasis is placed on operations to interdict and harass insurgent guerrilla units, training areas, and logistical installations and to deny insurgent forces access to local supply sources. These operations can be most successful and effective when Special Forces detachment personnel accompany long-range patrols and host country Special Forces units on deep penetrations into insurgent controlled territory. When properly coordinated with other receiving state activities conducted by regular forces (air strikes and major offensives against strongholds) these interdiction operations can make a significant contribution to the destruction of the insurgent threat. Major goals are to—

   (1) Destroy and damage supply routes and depots.
   (2) Capture equipment and key personnel.
   (3) Create confusion and weaken insurgent morale.
   (4) Force the insurgent to keep on the move.
   (5) Fragment the insurgent force.

(6) Relieve villages of the insurgent threat.
(7) Deny the insurgent the support of the local population.

b. Types of missions assigned paramilitary forces are basically the same as those conducted in guerrilla warfare operations; however, additional missions not normally associated with interdiction operations may include—
(1) Border operations (surveillance and denial).
(2) Reaction force operations.
(3) Reconnaissance and combat patrols.
(4) Long-range patrols into insurgent controlled areas.
(5) Psychological operations.
(6) Military civic actions.

c. Paramilitary forces, directed by Special Forces, conducting offensive and defensive operations against an insurgent force have certain advantages that are denied the guerrilla force in limited or general war. Some of these are—
(1) Artillery support from guns outside insurgent-controlled areas.
(2) Immediately available, close air and other air support.
(3) Reinforcements particularly from mobile airborne and ranger units.
(4) Evacuation from the operational area if necessary. Paramilitary forces, because of their location, organization, and support, can conduct extensive defensive operations in support of their villages and hamlets. Defensive tactics em-

ployed by paramilitary forces are similar to those of conventional forces with the exception of more primitive techniques for securing areas. These techniques include such defense measures as moats, palisade fences, man-traps, and terrain stripped of concealment and cover for an attacking force.

48. Border Operations

a. General. In a majority of cases where a subversive insurgency has been successful against an established government, support from sources outside the country has been a key factor in its success. A contributing factor to success in counterinsurgency is a denial of this external support which includes the use of adjacent countries as a sanctuary.

b. Border Control. Indigenous forces may be given missions of accomplishing varied border control operations concurrently with other military operations. There are two basic concepts of control—

(1) *Border denial.* These measures are taken to physically separate the insurgent force from external support provided from an adjacent country (see FM 31–10).

(2) *Border surveillance.* This consists of an extensive network of observation posts and watchers, augmented by intensive patrolling activity to detect, ambush, and

destroy small groups of infiltrators. Normally, only border surveillance is especially applicable to Special Forces-directed indigenous forces, operating in rugged terrain where construction of physical barriers is unfeasible. For a discussion of border control in counterguerrilla operations, see FM 31-16.

  c. Organization for Operations. The primary organization for border control operations is based on the insurgent situation and the terrain in the area. Special Forces commanders develop a force capable of sustained operations in remote areas for given periods of time. As a guide, a company of approximately 150 men can effectively control up to 10 kilometers of terrain of rugged hill masses and forests or jungles. Organizations are developed to insure adequate communications, fire support, and a highly trained reaction force as reserve.

  d. Bases of Operation. The Special Forces detachment commander in planning border control operations must consider—

  (1) *Range of fire support weapons.* If fire support from bases are provided by 105-mm and 155-mm howitzers, their respective ranges will permit location and deployment of base camps up to 30 kilometers apart. Shorter range weapons will not permit as much distance. A larger organization will be required to effectively control the assigned area.

(2) *Mobility of reaction force (reaction time)*. Maximum use of civilian transportation should be exploited; however, lack of fuel and other necessary POL may negate the use of civilian vehicles. Sufficient helicopter support must be available on a continuing basis in those areas which are most active.

(3) *Communications requirements*. The detachment commander must determine his communications needs and coordinate with the Special Forces group signal officer for procurement of additional signal equipment and of needed technical advice. Communications systems must be designed to tie in with existing systems and must include connection with local, intelligence gathering agencies and reaction and back-up forces. Indigenous personnel are trained in the use of any system established.

(4) *Effective span of control*. This is effected by an adequate communications system and strong, well-trained leaders. Conscientious advisory support from Special Forces personnel will insure maximum effort. Patrols and outposts will be given explicit directions to cover most contingencies that may arise and constant patrolling and periodic inspections of outposts and observation posts will further insure compliance with issued orders.

(5) *Logistical support of operational units.* The Special Forces commander and his counterpart, in committing long-range patrols into insurgent controlled areas or in pursuit of insurgent forces, must plan for a sound logistical resupply system. Lack of adequate supplies shorten the range capability of these patrols and render them ineffective. The detachment commander may consider using the area drop zone system for resupply. Procedures used will serve as a technique of control through the establishment of phases lines, so that he knows exactly where his patrols are at all times. Supplies also may be delivered to fixed outposts periodically. Changes of personnel and constant relocation of other than fixed outposts may negate such resupply; however, sufficient rations must be taken in to last for a given period of time. Consideration will be given to the use of pack animals where they are locally available, particularly in areas where predictable periods of poor flying weather, such as a monsoon season, is experienced.

e. Intelligence. The Special Forces operational detachment plans for the development of intelligence nets in the operational area to supply him a constant flow of intelligence. Military intelligence personnel, trained in agent recruiting and net organization are infiltrated into the operational area to develop the intelligence potential, and agents native to the area may be brought in by—

   (1) Preparing cover stories (i.e., merchants, farmers) for entry into populated areas.
   (2) Their assignment to paramilitary forces and local law enforcement agencies with freedom of movement throughout the area.

f. Operational Techniques. The techniques involved in border control operations are many and varied. They are limited only by the planning given to the operation by the detachment commander and his support elements. Techniques can include—

   (1) Saturation patrolling, to the maximum extent possible, with no fixed patterns and times.
   (2) Small-unit operations of squad size or below.
   (3) Surveillance of insurgent activities from fixed locations by day, with active operations against targets of opportunity during periods of darkness.
   (4) Deceptive measures taken when moving forces through the area of deceive the insurgent forces in their surveillance activities.
      (*a*) Use of civilian clothing to hide uniforms and identity, e.g., dress patrols as farmers or workers so that they

can move freely through an area without suspicion.
- (b) Use of enemy clothing when traveling over trails normally used by the insurgent.
- (5) Penetration of the insurgent force and their support elements, by selecting and training local indigenous personnel.
- (6) Infiltration of homing devices into the enemy organization by allowing them to capture arms, ammunition, radios, or other equipment essential to their activities.
- (7) Gathering intelligence by using—
  - (a) A centralized system for maintaining records on prisoners, suspects, or criminals, who, through the promise of parole or pardon, may be used to solicit information or act as agents or informers. The inherent risk, of course, is the possibility of double agents. Also the detachment commander can successfully employ deception operations by implanting false information, or other means.
  - (b) Intelligence maps which indicate routes, movement times, and primary and alternate base locations, which, when cross-referenced with other files, will indicate the tentative activities of the insurgent in the area, substantiation by surveillance of the popula-

tion, and attitudes and reaction to curfew and other restrictions. For additional information, FM 31–10, FM 31–16, FM 31–21, FM 31–21A, and FM 31–73.

Section III. TRAINING OF INDIGENOUS FORCES

49. General

Before commitment into operational areas, Special Forces detachments simplify their task of training the indigenous forces by developing a tentative training program. Guided by area studies and intelligence, the detachment prepares and collects training aids and other equipment that may be required in the operational area. When committed into a GWOA, these items are delivered with the detachment's automatic supply drop.

50. Considerations

The following factors dictate whether a centralized, decentralized, or a combination of these training systems will be the most effective method of training an indigenous force:

a. Mission assigned.

b. Enemy or insurgent capability.

c. Level of training and operational readiness of the indigenous force.

d. Time available.

e. Facilities and equipment available.

f. Climate and terrain in the area.

51. The Training Program

After the detachment has entered the area, several factors are considered in developing the training program for the indigenous forces—

a. Selection of Essential Subjects. These are determined by evaluating the operational mission, the training and equipment status of the indigenous forces, the enemy or insurgent situation, and the population's attitude and logical support.

b. Organization and Training. In each situation, the detachment commander decides which of the following training system will be the most beneficial to the indigenous force:

   (1) Individual or on-the-job training.
   (2) Centralized or decentralized training.
   (3) Specialized schools for selected personnel.

Examples of a master training program for a leadership school and a 30-day master training program for preparing individual training are shown in appendix VIII. Time for each training phase is established by the detachment's mission.

c. Preparation of training aids and facilities. Each member of the detachment contributes his ideas and thoughts in the preparation of training aids and facilities. Improvisations and use of local craftmen aid materially in the production of training aids and facilities.

d. Administration. Training is planned, supervised, and inspected by detachment members and their counterparts. Each area of training is supervised by detachment members while indigenous personnel conduct training.

52. Training Operations

a. Unconventional Warfare. Throughout the organization, development, and training phases of guerrilla activities, combat operations are conducted. With the aid of prior and concurrent psychological operations, the goals of these are to—
  (1) Succeed in attracting additional recruits to the guerrilla force.
  (2) Assist in gaining support of civilian population.
  (3) Give the area command an opportunity to evaluate the training conducted.
  (4) Increase the morale and esprit of the guerrilla force.

b. Counterinsurgency Operations. Primary consideration is given to training local defense units in defensive operations and in using weapons and equipment for security. Throughout the training program constant attention is given to the psychological preparation of the people to accept government support in construction programs, and in establishing sound local government. During the training programs, combat operations are conducted on a limited scale as required to rid the area of insurgent threats and activities. These operations are controlled and directed by Special Forces personnel.

CHAPTER 6

AIR OPERATIONS

Section I. GENERAL

53. Responsibilities in Unconventional Warfare/Counterinsurgency Operations

In conventional military operations the selection of DZ's or LZ's is a joint responsibility of both the Air Force and the Army. The marking of these sites for identification purposes is the responsibility of the Air Force. The nature of Special Forces in guerrilla warfare and their capabilities for extended operations in remote areas in support of counterinsurgency operations requires Special Forces to assume the responsibility for these functions. Special Forces personnel are trained in procedures for selecting, reporting, and marking DZ's and LZ's and for organizing and conducting reception operations.

a. Before Infiltration. Special Forces operational detachments should select DZ's and LZ's before infiltration by using all available intelligence sources and available maps. The DZ and LZ data are then filed at the SFOB. After infiltration and upon completion of ground reconnaissance, the detachment then either confirms or makes line changes to the data on file, thereby limiting the volume of radio traffic pertaining to air support information. Final approval of infiltration DZ's and LZ's is a joint decision of the commanders of the SFOB and the air support unit.

b. After Infiltration. Following commitment into operational areas, Special Forces detachments are responsible for selecting, reporting, and marking DZ's and LZ's. Final acceptability of the DZ rests with the air unit performing the mission.

54. Air Delivery Operations

a. Characteristics. Air delivery operations in support of Special Forces units are characterized by—
  (1) Single aircraft missions.
  (2) Penetration flights into denied areas under conditions of limited visibility and at varying flight levels (to include low level flying at 500 feet or below, and high altitude flights in excess of 33,000 feet for HALO operations).
  (3) Frequent changes of course enroute to the initial point (IP).
  (4) Departure from IP on a predetermined track.
  (5) Arrival over DZ within a specified time block.
  (6) Execution of the drop directly over the DZ ground release point markings.

(7) Drops conducted at altitudes between 400–800 feet. Cargo drops in selected areas, supporting counterinsurgency operations, may be conducted successfully at altitudes of 250 to 350 feet.

(8) A single pass over the DZ.

(9) Maintaining drop course, altitude, and speed until at least 47 kilometers away from the DZ.

(10) The use of alternate DZ's.

(11) The use of blind-drop procedures when reception committees are unavailable.

b. Air Operations in Operational Areas. Successful air delivery operations depends upon careful coordination between the operational detachment and the air support unit and compliance with SOP's. Coordination is accomplished through the SFOB (fig. 1). A typical air resupply mission involves the following sequence of events:

(1) *Operational detachment.*
   (a) Selects DZ's and LZ's.
   (b) Transmits DZ or LZ data and resupply requests to SFOB.

(2) *SFOB.*
   (a) Processes DZ, LZ, data, and resupply requests.
   (b) Coordinates mission with air support unit.
   (c) Transmits mission confirmation message to operational detachment.

(d) Prepares and delivers supplies and personnel to departure airfield.

(3) *Air support unit.*

(a) Prepares mission confirmation data for SFOB.

(b) Receives and loads supplies and personnel to be delivered.

(c) Executes mission.

(4) *Operational detachment.*

(a) Organizes and receives personnel and supplies.

(b) Distributes incoming supplies.

c. Packaging. Supplies are packed and rigged in aerial delivery containers which have a capacity of 230 kg or less. To facilitate rapid clearance of the DZ, the contents of each container are further packaged in man-portable increments. Detailed information on types and uses of various containers are found in TM 57–210 and special texts published by the U.S. Army Quartermaster School, Fort Lee, Va. 23801; also see Catalog Supply System, appendix IV, and chapter 10.

55. Preplanned Resupply

a. Automatic Resupply. Before infiltration, plans and coordination are made for an automatic resupply mission to be flown for the operational detachment on a specific date following infiltration. Delivery is made on a DZ selected through map reconnaissance and available intelligence. Primary and alternate DZ's are selected (ch. 10). Before arrival of the automatic resupply, this mission may be modified or canceled. Automatic resupply operations are also adaptable to support counterinsurgency and counterguerrilla operations. DZ's previously selected may be used as phase lines and resupply for extended operations of strike forces and elements in pursuit of an insurgent force.

b. Emergency Resupply. Following infiltration a detachment selects a DZ for emergency uses and reports to the SFOB as soon as feasible. The location of the DZ will not be known to the guerrillas. It will be located away from anticipated enemy and guerrilla activities. The SOI/SSI indicates procedures to be followed to put the emergency resupply operations into effect. As an example the SOI/SSI may indicate that if the detachment fails to make a prescribed number of scheduled radio contacts, then after a specified number of days following the last scheduled contact the emergency resupply will be flown. The marking pattern for the emergency resupply mission is coordinated before infiltration; but the mission is flown, and the supplies dropped on the DZ even though no markings are visible. This blind drop technique is used in the event the detachment, or surviving individuals, are under such enemy pressure that arrival at the DZ in time to receive the drop or to risk identification of the DZ is precluded (ch 10).

Section II. DZ SELECTION AND REPORTING

56. Criteria

The selection of a DZ must satisfy the requirements of both the aircrew and the reception committee. The aircrew must be able to locate and identify the DZ. The reception committee selects a site that is accessible, reasonably secure, and safe for delivery of incoming personnel or supplies.

  a. Air Considerations.
  (1) *Terrain.*
    (*a*) The general area surrounding the site must be relatively free from obstacles which may interfere with safe flight.
    (*b*) Flat or rolling terrain is desirable; however, in mountainous or hilly country, sites selected at higher elevations such as broad ridges and level plateaus can be used.

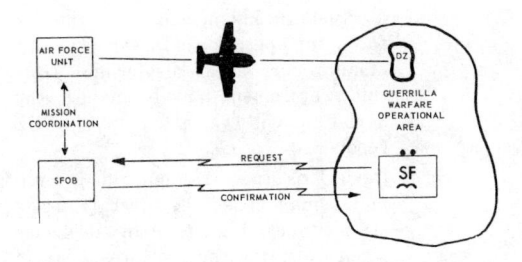

Figure 1. Coordination of air operations.

(c) Small valleys or pockets completely surrounded by hills are difficult to locate and normally should not be used.

(d) In order to afford the air support unit flexibility in selecting an IP it is desirable that the aircraft be able to approach the center of the DZ from any direction. It is desirable that there be an open approach quadrant of at least 90° to allow the aircrew a choice when determining their approach track.

(e) DZ's having a single clear line of approach are acceptable, provided there is a level turning radius of 5 kilometers on each side of the site (1.5 kilometers for light aircraft (fig 2).

(f) Rising ground or hills of more than 305 meters of elevation above the surface of the site normally should be at least 16 kilometers from the DZ for night operations. In exceptionally mountainous areas deviations from this requirement may be made. Any deviation will be noted in the DZ report.

(g) Deviations from recommended minimums may cause the aircraft to fly at altitudes higher than desirable when executing the drop, resulting in excessive wind drift.

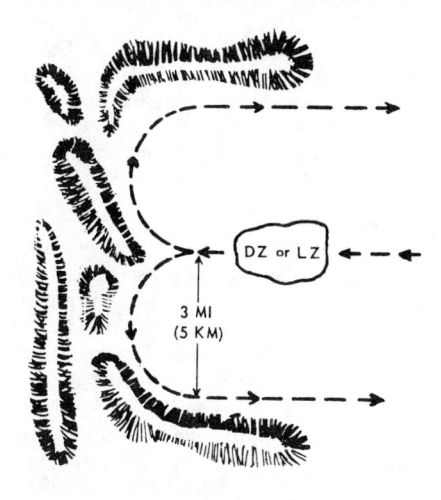

Figure 2. Level turning radius required for one approach DZ's and LZ's (medium aircraft).

- (2) *Weather.* The prevailing weather conditions in the drop area must be considered. Ground fog, mists, haze, smoke, and low-hanging cloud conditions may interfere with visual signals and DZ markings. Excessive winds also hinder operations.

- (3) *Obstacles.* Due to the low altitudes at which optional drops are conducted, consideration must be given to navigational obstacles in excess of 90 meters above the level of the DZ and within a radius of 8 kilometers. When operational drops are scheduled for altitudes less than 400 feet, specific considerations should be given to navigational obstacles in excess of 30 meters. If such obstacles exist and are not shown on the issued maps, they must be reported.

(4) *Enemy concentration.*
   (*a*) Because of aircraft vulnerability in guerrilla warfare operational areas, drop site locations should preclude the aircraft's flying over or near enemy air installations or known concentrations in the final approach to the DZ as well as the departing track.
   (*b*) In counterinsurgency operations outer security should conduct constant surveillance and patrolling to prohibit the enemy from positioning automatic small arms in the area surrounding the DZ. During its landing roll and take-off run, the aircraft should be flanked and paced by machinegun vehicles prepared to deliver saturation fires to either flank of the runway.

b. Ground Considerations.

(1) *Shape and size.*
   (*a*) The most desirable shapes for a DZ square or round. This permits a wider choice for the aircraft approach track.
   (*b*) **The required length of a DZ depends primarily on the number of units to**

be dropped and the length of their dispersal pattern.

1. Dispersion occurs when two or more personnel or containers are released consecutively from an aircraft in flight. The long axis of the landing pattern generally parallels the direction of the flight (fig. 3).

2. Dispersion is computed using this rule-of-thumb formula—$\frac{1}{2}$ speed of aircraft (knots) $\times$ exit time (seconds) = dispersion (meters). Exit time is the elapsed time between the exits of the first and last items.

3. The length of the dispersion pattern represents the absolute minimum length required for DZ's. If personnel are to be dropped, a safety factor of at least 100 meters is added to each end of the dispersion pattern to ascertain minimum DZ length required.

(c) The width of rectangular-shaped DZ's should allow for minor errors in computation of wind drift.

(d) The size of a DZ will be dictated by its use. For personnel drops the dispersion pattern, wind drift, and the cleared area required for the display of the DZ markers will be considered. When dropping personnel, use a DZ measuring at least 300 by 300 meters.

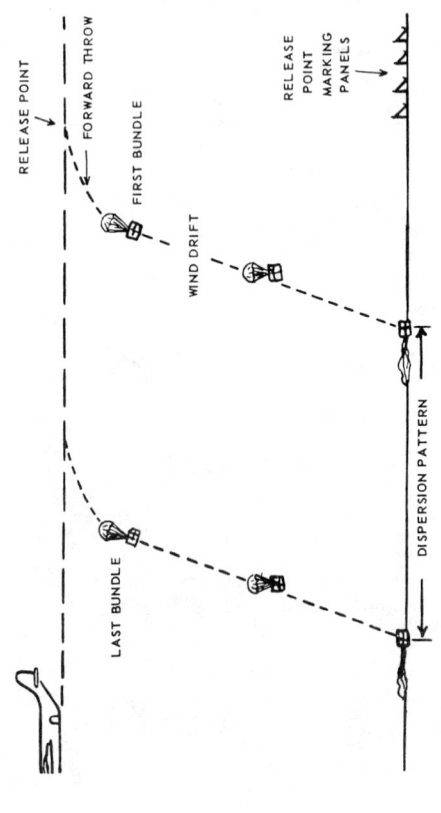

Figure 3. Computation of dispersion.

(2) *Surface.*
  (a) The surface of the DZ should be reasonably level and free from obstructions such as rocks, trees, fences, and powerlines. Tundra and pastures are ideal types of terrain for both pernel and cargo reception.

  (b) Personnel DZ's located at comparatively high elevations (1,840 meters or higher) will, where possible, use soft snow or grasslands. Because of the increased rate of parachute descent at these altitudes, such drops are less desirable than those at or near sea level.

  (c) Swamps, paddies, and marshy ground are marginally suitable for personnel and bundles in the wet season and for bundles when frozen or dry. The presence of water compounds the recovery problems and is hazardous for personnel. Frozen paddies present a rough, hard surface, marginally suitable for personnel drops.

  (d) Personnel and cargo can be received on water DZ's.
    1. In dropping personnel on a water DZ, the depth will not be less than 1.2 meters and arrangement must be made for rapid pickup.
    2. The surface of the water will be clear of floating debris or moored craft;

and there will be no protruding boulders, ledges, or pilings.

3. The water will also be clear of underwater obstructions to a depth of 1.2 meters.
4. Water reception points will not be near shallows or where currents are swift.
5. Minimum safe water temperature is +50 degrees F (+10°C).

(e) The following ground surfaces can be used for supply drop zones.
1. Gravel or small stones no larger than a man's fist.
2. Agricultural ground; however, if post mission secrecy is a factor, it is inadvisable to use cultivated fields.
3. Brush or tall trees; however, marking of the DZ and the recovery of containers is more difficult.
4. Marsh, swamp, or water provided the depth of water or growth of vegetation will not result in loss of containers.

(3) *Security.* Special Forces operations makes security a matter of prime importance. The basic considerations for security in the selection of DZ's are—
(a) Location to permit maximum freedom from enemy ground interference.
(b) Accessibility to the reception committee by routes that are concealed from

enemy observation or which can be secured against interdiction or ambush.
- (c) Nearness to areas suitable for the caching of supplies and disposition of air delivery equipment.

57. Reporting Drop Zones

a. Drop Zone Data. The minimum required drop zone data includes—
- (1) Code name extracted from the operational detachment SOI. Also indicate if DZ is primary, alternate, or water.
- (2) Location complete with military grid coordinates of the center of the DZ.
- (3) Open quadrants measured from the center of the DZ, reported as a series of magnetic azimuths clockwise from the north. The open quadrant delineates acceptable aircraft approaches (fig. 4).
- (4) Track with magnetic azimuth of recommended aircraft approach (fig. 4). If a specific aircraft course is required it will be reported as "required track." Only exceptional circumstances will cause the detachment to require a specific track to be flown.
- (5) Obstacles to flight over 90 meters in elevation above the level of the DZ, within a radius of 8 kilometers, not shown on the issued maps of the areas. Obstacles are reported by description, magnetic azimuth, and distance from the center of the DZ (fig. 5).

(6) Reference point such as a landmark that can be located on issued maps by name alone; e.g., a lake, town, or mountain. It is reported by name, magnetic azimuth, and distance from the center of the DZ to the center of the reference. It is used with (2) above in verifying the DZ location and it should not be confused with the initial point (IP) selected by the aircrew.

(7) See appendix III for sample drop zone report.

b. *Additional Items.*

(1) *Concurrent mission request.* The basic drop zone report may become a mission request by the addition of two items.

   (a) Date time group that indicates the actual time that the aircraft is desired over the DZ. Use Greenwich Mean Time (ZULU).

   (b) Items or services desired. Requests for supplies are normally extracted from the Catalog Supply System (CSS) established in the unit SOP.

(2) *Designation of alternate DZ.* When a concurrent mission request is submitted with the DZ report, an alternate DZ should be designated. The code name of the alternate DZ is the last item of the mission request, if a DZ has been previously reported. If the alternate DZ has not been previously reported, then the

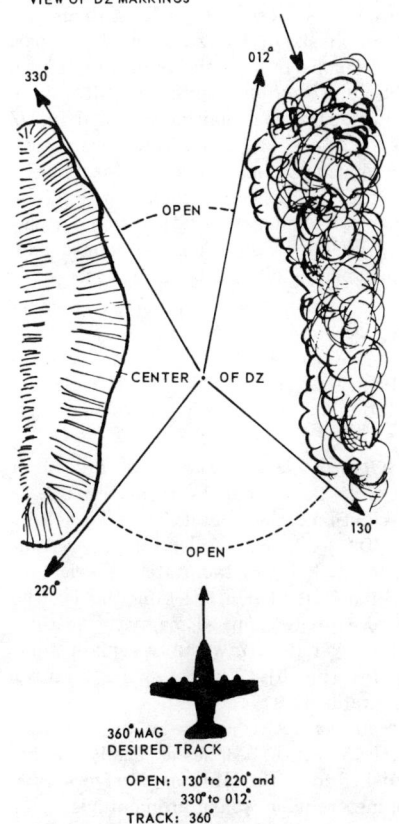

TREES OR TERRAIN THAT WOULD MASK PILOT'S
VIEW OF DZ MARKINGS

360° MAG
DESIRED TRACK

OPEN: 130° to 220° and
330° to 012°.
TRACK: 360°

Figure 4. Computation of open quadrant and desired heading.

mission request will include items one through six of the standard DZ report as they pertain to the alternate DZ.

(3) *Special situations.* In special situations, additional items may be included in DZ reports; e.g., additional reference points, navigational check points in the vicinity of the DZ, special recognition, and authentication means. If additional items are included they must be identified by appropriate paragraph headings.

c. Azimuths. Azimuths are reported as magnetic in degrees and in three digits. With the exception of the aircraft track, all azimuths are measured from the center of the DZ. For clarity, the abbreviation DEG will be used when reporting direction.

d. Initial Points (IP's). The IP, located at a distance of 8 to 24 kilometers from the DZ, is the final navigational checkpoint. The pilot selects the IP; the DZ party can assist him by recommending a track that will facilitate the selection of an identifiable IP. Upon reaching the IP, the pilot turns to a predetermined magnetic heading that takes him over the DZ within a certain number of minutes (fig. 6). The following features constitute suitable IP's:

(1) *Coastlines.* A coastline with breaking surf or white beaches is easily distinguished at night. Mouths of rivers over 50 meters wide, sharp promontories, and inlets are excellent guides for both day and night.

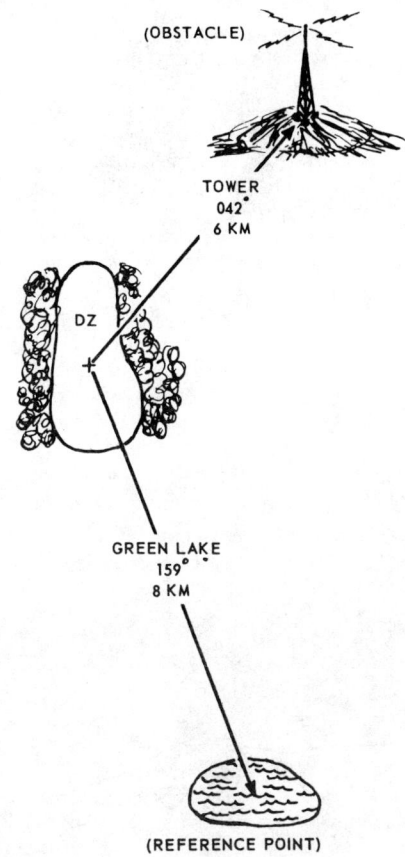

Figure 5. Reporting of obstacles and reference points.

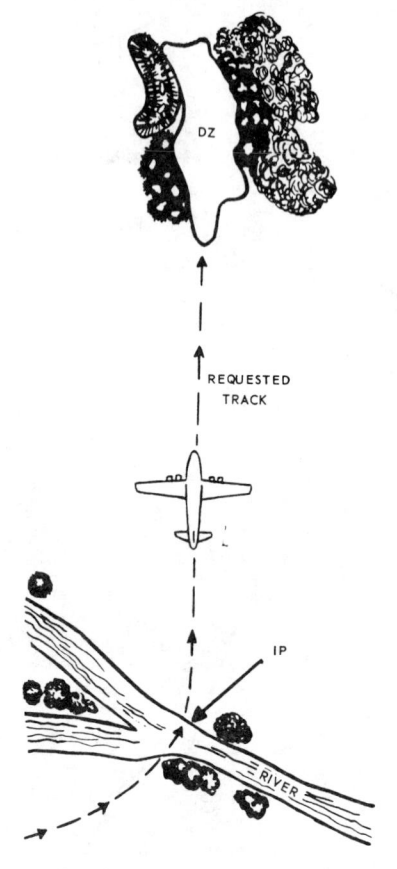

Figure 6. Relationship between IP and requested track.

(2) *Rivers and Canals.* Wooded banks reduce reflections, but rivers more than 30 meters wide are visible from the air. Canals are easily recognizable because of their straight banks and uniform width; however, they may be valueless in areas where they are uniformly patterned.

(3) *Lakes.* Lakes, at least 1 square kilometer in size, give good light reflection but must be clearly identifiable because of shape or some other distinctive characteristics.

(4) *Forest and woodlawns.* Forested areas of at least one-half kilometer square with clearly defined boundaries or unmistakable shapes are easily identified.

(5) *Major roads and highways.* Straight stretches of main roads with one or more intersections can be used as an IP. For night recognition, dark surfaced roads are not desirable; although, when the roads are wet reflection from moonlight is visible.

(6) *Railways.* When there is snow on the ground, rail lines which are frequently used will appear as black lines cutting through the white landscape. Under other than snow conditions, where rail lines make junction with other prominent land marks, such as highways,

bridges and tunnels, they can be used as IP's.

e. Subsequent Use of DZ. SFOB maintains reported DZ's on file. If a previously reported DZ is to be used again, the mission request need contain only—

(1) Code name of DZ.
(2) Date/time mission.
(3) Supplies/services desired.
(4) Alternate DZ.
(5) Track of alternate DZ.

58. Alternate Drop Zones

To increase the probability of success in receiving supplies and personnel, an alternate DZ will be designated for every mission requested. Separate drop times are established and both DZ's will be manned by at least a skeleton reception committee. If the primary DZ is not suitable for reception due to unfavorable conditions, the aircraft proceeds to the alternate DZ. Drop times for alternate DZ's are determined by the air support unit and are included in the mission confirmation message (app. III).

59. Mission Confirmation for Air Drop

Following the processing of the DZ report and resupply request at SFOB, a confirmation message is transmitted to the operational detachment usually by blind transmission broadcast (BTB). The confirmation message includes—

a. Code Name of the Drop Zone. Code name identifies the mission.

b. Track. The magnetic azimuth upon which the aircraft will approach. The actual track flown may differ from the original request to conform to the location of the IP selected by the aircrew.

c. Date Time of Drop. This also may differ from the original request because of priorities, weather, or aircraft availability.

d. Number of Cargo Containers or Personnel. This assists the reception committee in the recovery of all containers and personnel.

e. Drop Altitude. This assists the reception committee in properly placing DZ release point markings to compensate for winddrift.

f. Alternate DZ. See appendix III for sample drop confirmation messages.

g. Drop at Alternate. Date/Time of drop at alternate DZ.

60. Area Drop Zone

a. General. An area DZ consists of a prearranged flight over a series of acceptable drop sites which establishes a line of flight between two points, (A and B, fig. 7). The distance between these points should not exceed 25 kilometers, and will have no major changes in ground elevation in excess of 90 kilometers. Drop sites may be selected not more than 1 kilometer to the left or right of the established line of flight.

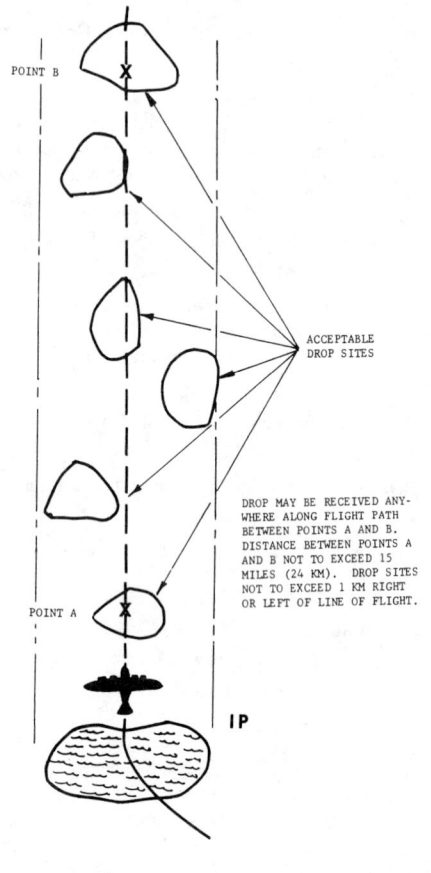

Figure 7. Area drop zone.

The aircraft arrives at point A at the scheduled time and proceeds towards point B looking for the DZ markings. Once the markings are located, the drop is conducted in the normal manner. The area DZ system is particularly well adapted for use in conjunction with preplanned automatic resupply operations where DZ's are frequently selected on the basis of map reconnaissance. The area DZ system is also adaptable to long-range patrols and operations conducted by counterinsurgency elements in pursuit of an insurgent force.

b. Drop Zone Data. Area DZ's are reported using the normal DZ report format, with these exceptions.

  (1) Locations of both point A and B, including reference points.

  (2) The open quadrant is reported as "none."

  (3) Obstacles over 90 meters above the level of the terrain along the line of flight, and within 8 kilometers on either side and not shown on the issued map (fig. 8). These obstacles are reported in reference to either point A or B.

  (4) See appendix III for sample area DZ report.

Section III. MARKING DROP ZONES

61. Drop Zone Identification

The purpose of DZ markings is to identify the

site for the aircrew, indicate the point over which personnel or cargo will be released, and provide

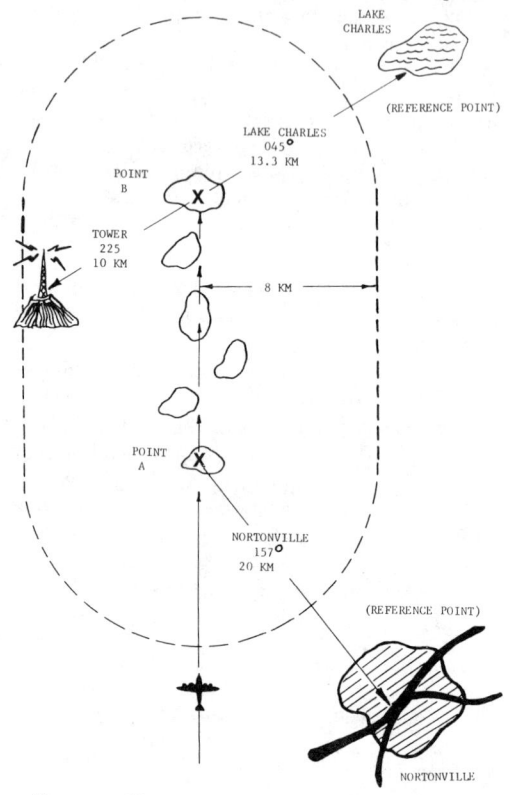

Figure 8. Obstacles and reference points (area DZ).

a visual track for the aircraft. The procedures for marking DZ's are included in the SOI.

  a. Nightime marking of DZ's is accomplished by using lighting devices such as flashlights, flares, and small wood, oil, or gas fires.

  b. For daylight operations a satisfactory marking method is the Panel Marking Set AP-50 or VS-16. If panels are not available, sheets, strips of colored cloth, or other substitutes can be used provided there is a sharp contrast with the background. Smoke grenades or simple smudge fires used in conjunction with other markings greatly assist the aircrew in sighting the DZ markings on the approach.

62. Homing Devices

The use of electronic homing devices permits reception operations during conditions of low visibility. Such devices may also be used in conjunction with visual marking systems.

63. Computation of Ground Release Point

  a. General. The release point is determined to insure delivery of personnel or cargo within the usable limits of the DZ. Computation of the ground release point involves the factors in *b* through *d* below.

  b. Personnel and Low Velocity Cargo Drops.
   (1) *Dispersion.* Dispersion is the length of the pattern formed by the exit of the parachutists or cargo containers (fig.

3). The desired point of impact for the first parachutist container depends upon the manner in which the calculated dispersion pattern is fitted into available DZ space.

(2) *Wind drift.* This is the horizontal distance traveled from the point of parachute deployment to the point of landing as a result of wind conditions. The release point is located a calculated distance upwind from the desired impact point. To determine the amount of drift, use the following rule of thumb formulas:

(a) For personnel using the T-10 parachute: Drift (meters) = altitude (hundreds of feet) × wind velocity (knots) 4 (constant factor).

(b) For all other low velocity parachute drops: Same as 1 above; however, instead of 4 use the constant factor 3.

(c) For a mixed load of personnel and cargo, use the personnel factor (4).

(3) *Forward throw.* This is the horizontal distance traveled by the parachutist or cargo container between the point of exit and the opening of the parachute. This factor, combined with reaction time of the personnel in the aircraft, is compensated for by moving the release point an additional 100 meters in the direction of the aircraft approach.

(4) *High Velocity and Free-Drop Loads.* High velocity and free-drop loads are not materially affected by wind conditions; therefore, wind drift is disregarded. Dispersion is computed the same as for low velocity drops. On the other

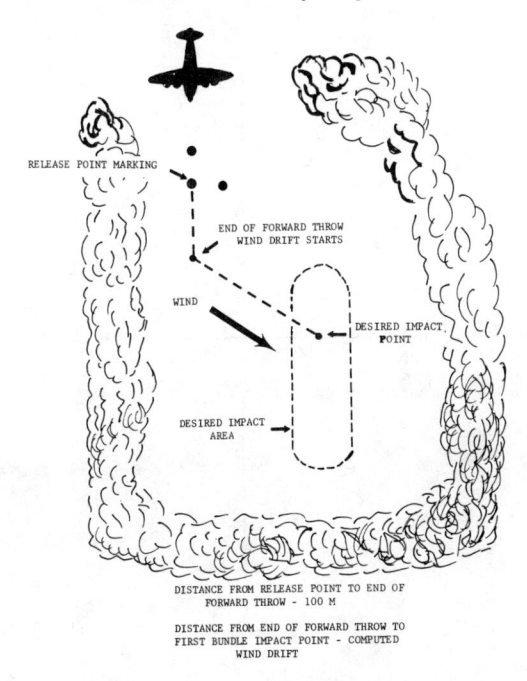

DISTANCE FROM RELEASE POINT TO END OF
FORWARD THROW - 100 M

DISTANCE FROM END OF FORWARD THROW TO
FIRST BUNDLE IMPACT POINT - COMPUTED
WIND DRIFT

Figure 9. Computation of release point.

hand, without the restraint of a parachute, forward throw is compensated for by moving the ground release point marking in the direction from which the aircraft will approach; a distance equal to the altitude of the aircraft above the ground.

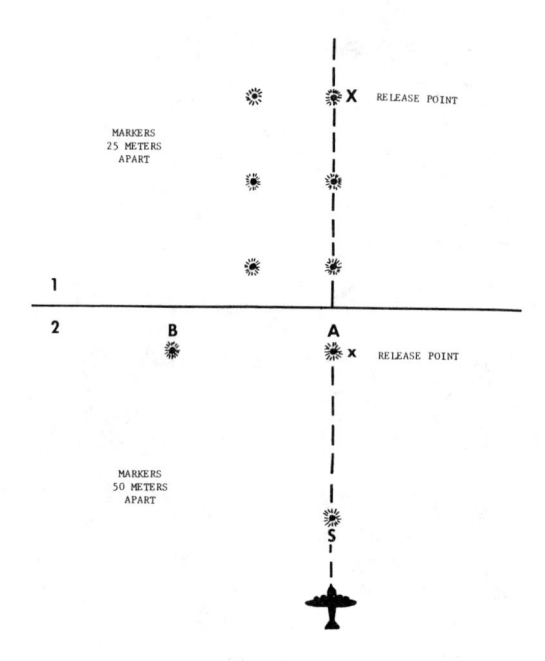

Figure 10. Methods of release point marking.

64. Methods of Release Point Marking

There are two methods for marking the DZ release point. The principal difference between the two is the method of providing identification. The marking systems described in *a* through *c* below are designed primarily for operational drops executed at an altitude of 400 to 800 feet. Training jumps executed at an altitude in excess of 800 feet require modification of the marking systems.

a. The primary marking method employs lights or panels in a distinctive configuration which changes daily according to unit SOI. In addition to marking the ground release point this configuration serves to identify the drop zone to the aircrew.

  (1) The number of markers used seldom exceeds six.

  (2) The distance between markers is 25 meters for drops executed at operational altitudes of 800 feet or less. When the drop altitude exceeds 800 feet, the spacing is increased by 50 meters.

  (3) The release point markers normally will form a distinctive shape (square, rectangle, triangle) or letter (T, L, X).

  (4) In executing drops, the aircraft is alined as accurately as possible over the right hand row of markers. Deviations will not exceed 50 meters to the right of the row of markers. The drop is made

directly over the last light in the right hand row (fig. 10).

b. The alternate marking method employs a standard three-marker pattern (fig. 10) positioned in the form of an inverted L. Identification of the drop zone is accomplished by means of a code light displayed in addition to the 3 lights of the standard inverted L, placed 5 meters from the stern marker S.

c. All jumps from an absolute altitude in excess of 800 feet require the use of a flank marker placed 200 meters to the left of the release point markings. The configuration of some cargo and troop carrying aircraft prevent the pilot from seeing the markings after approaching within approximately 1.6 kilometers of the DZ while flying at 1,000 feet absolute altitude. From this point on, the pilot must depend on flying the proper track in order to pass over the release point. The flank marker indicates the release point and the exact moment the drop should be executed. Operational drops executed at altitude less than 800 feet do not require the flank marker.

65. Placement of Markings

a. Markings must be clearly visible to the pilot of the approaching aircraft. The formula for determining mask clearance is 15 units of horizontal distance, a ratio of 15 to 1, for each unit of obstruction. As an example, markings must have a clearance of at least 450 meters from an obstruction 30 meters in height (fig. 11).

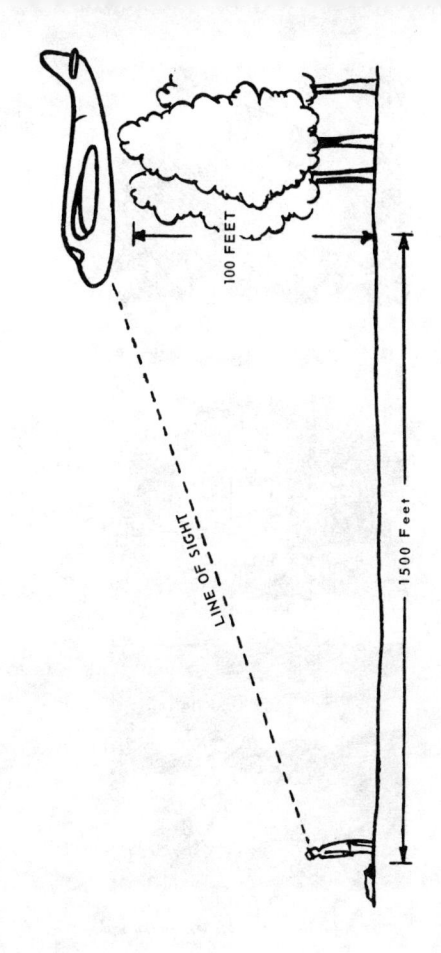

Figure 11. MASK clearance RATIO 15:1.

FLASHLIGHT HOOD

FIRE SCREEN

FIRE PIT

Figure 12. Security of DZ markings.

b. Additionally, precautions must be taken to insure that the markings can be seen only from the direction of the aircraft approach. Flashlights are sufficiently directional not to require shielding if aimed toward the flight path. Fires or improvised flares are screened on three sides or placed in pits with sides sloping toward the direction of the aircraft's approach (fig. 12).

c. When panels are used for daylight DZ marking they are positioned at an angle of approximately 45° from the horizontal to present the maximum surface toward the approaching aircraft (fig. 13).

66. Unmarked Drop Zones

a. Personnel and equipment may be dropped on unmarked drop zones when necessary. This technique is generally limited by visibility to specific moon phases or daylight. A drop zone selected for this purpose should be located in an isolated or remote area and free from enemy interference.

b. Drops on unmarked DZ's may be preplanned for specific periods of time. The receiving unit is required to keep the DZ under constant surveillance during the time the drop is scheduled. As soon as the cargo is delivered, observers alert the receiving unit, measures are taken to dispose of the items received, and the DZ is sterilized (obliterating all signs of the drop). To aid in recognition, the DZ's should be of odd configuration and size and have specific, recognizable land marks. Electronic signaling devices, compatible with the aircraft equipment, should be used with a previously planned recognition signal where possible.

   c. On the basis of available intelligence and

Figure 13. Placement of panel markings.

information from the operational area concerning weather conditions and prevailing winds, the pilot will compute the air release point.

Section IV. HALO OPERATIONS

67. General

When atmospheric conditions over the target area prevent the infiltrating groups from sighting the DZ as they exit the aircraft, special techniques are employed. If electronic equipment compatible with aircraft radio and radar facilities is available and used, the time of exit may be determined by a ground release point system. If a trained reception committee with special communications equipment is not available, a computed air release point system may be used. In all cases, air crews are responsible for the release point.

68. DZ Markings

The visual ground marking release system is used when the aircraft can fly to the desired DZ with good forward ground visibility.

a. In HALO operations DZ markings indicate the impact point, not the release point. Markings will indicate wind direction and speed; flares, gas pots for night operations, and panel marking devices for daylight operations are set in the form of an arrow pointed into the wind (fig. 14).

b. Five marking devices placed at 25 meter intervals outline the arrowhead.

c. The stem of the arrow will act as the wind speed indicator, with markers placed at 25 meter intervals. The arrowhead indicates to jumpers 0 to 5 knots of wind; the first marker below the arrowhead, 5 to 10 knots; the second marker, 10 to 15 knots; and the third marker, 15 to 20

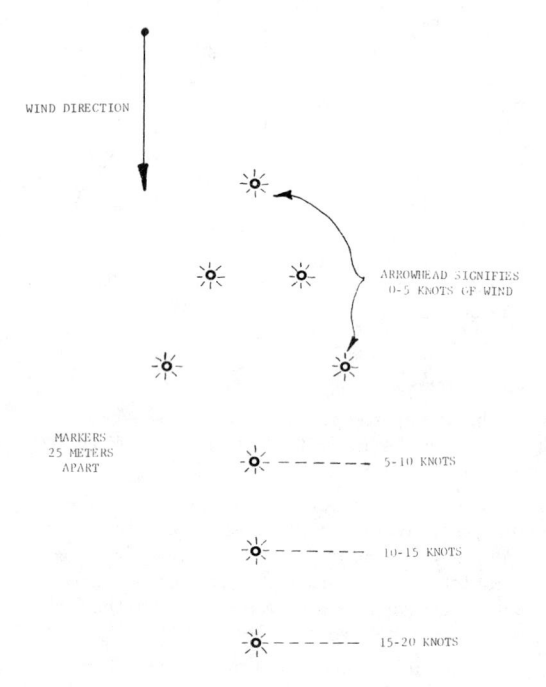

Figure 14. HALO DZ markings.

knots of wind. HALO parachutists, knowing their opening altitude, estimated wind speed, and direction as indicated by the markings on the ground, will mentally compute their wind drift after opening. Using the "tracking" technique for horizontal movement, they will move to their selected opening point. Additionally, visible markings on their designated leader enables them to assemble in the air while maneuvering for their opening.

d. The procedures for indicating an abort or on-jump status are the same as those outlined in this chapter. If DZ markings can be observed before the jump and wind indicators reflect winds in excess of 20 knots, there is no jump. Airborne operations orders will indicate rescheduling of the mission or cancellation.

69. Obscured DZ's

Situations often require the use of sites that are difficult to locate and identify. Atmospheric conditions may obscure the pilot's view of the ground. Then audio, radio, or radar devices must be used in conjunction with standard signals.

a. If electronic equipment is available for use, the electronic ground mark-release system may be used. This system requires personnel on the ground to place an electronic device in such a position that it can be interrogated by incoming aircraft. The device can be either passive or **active and will be selected** by the Air Force to **insure compatibility with** equipment installed in aircraft. When employing this system, good forward visibility from the aircraft is not required. The aircraft must be flown at a predetermined track, altitude, and air speed from the IP to the electronic device. When the aircraft is directly over the device the parachutists are released.

b. When no ground personnel are available to provide terminal air markings, the computed air release point system (CARP) is used. The aircrew must know the desired impact point and be able to visually identify landmarks short of the impact point and in the line of flight from the IP. The CARP system is the parachute release system used during normal, joint airborne operations.

70. Premission Planning

Joint premission planning will include the establishment of oxygen procedures, exit door opening procedures, magnetic course from IP, airspeed and altitude at release point, the release point, and recognition and abort procedures.

71. Resupply

Resupply procedures may be carried out by HALO techniques using various methods to free fall equipment into designated areas.

a. Nonelectric blasting caps may be used to fire parachute retaining devices and time fuse can be cut in accordance with altitude and desired free fall time of equipment.

b. Power actuated reefing line cutters, an item of issue available to airborne units, may be used when longer delays are necessary.

c. Use of high altitude bombing techniques are satisfactory for delivery of time-delay cargo parachutes.

Section V. DZ LANDING OPERATIONS

72. Reception Committees

It is desirable that the Special Forces detachment be met upon infiltration by an indigenous reception committee. Infiltration without the assistance of the reception committee may be necessary when there is no prior contact with the resistance element in the area and time does not permit the establishment of such contact. Once the detachment is within the operational area, reception committees are formed to conduct all future airborne or air landed operations within operational areas. Reception committees normally are composed of indigenous personnel trained and supervised by members of the Special Forces operational detachment. The functions of a reception committee are to—

a. Provide security for the reception operation.

b. Emplace DZ markings and air ground identification equipment.

c. Maintain surveillance of the site before and after the reception operation.

d. Recover incoming personnel and cargo.

e. Sterilize the site.

f. Provide for movement from DZ and disposal of equipment.

73. Composition and Duties

a. Organization of Reception Committee. The reception committee normally is organized into five parties—

- (1) Command party.
- (2) Marking party.
- (3) Security party.
- (4) Recovery party.
- (5) Transport party.

b. Command Party.

- (1) Controls and coordinates the actions of all reception committee components.
- (2) Includes the Reception Committee Leader (RCL) and communications personnel, consisting of messengers, a radio operator, and the Special Forces advisor.
- (3) Provides medical support during personnel drops.

c. Marking Party.

- (1) Sets up and operates the marking system.
- (2) Lights and extinguishes lights as directed.
- (3) Assists in recovery of personnel and equipment.

(4) Helps sterilize DZ, by covering all traces of light pattern.

d. *Security Party.*

(1) Insures that unfriendly elements do not interfere with the operation.

(2) Consists normally of inner and outer security elements.

   (*a*) The inner security element is positioned in the immediate vicinity of the site and is prepared to fight delaying or holding actions.

   (*b*) The outer security element consists of outposts established along approaches to the area. They may prepare ambushes and road blocks to prevent enemy movement toward the site.

(3) The security party will be supplemented by auxiliary personnel depending upon the operational environment in guerrilla operational areas and by local, self-defense units or civil defense groups in a counterinsurgency environment. These groups generally are used to maintain surveillance over enemy activities; keep the security party informed of enemy movements; and, when necessary, conduct limited objective attacks or ambushes to prevent enemy movement toward the site.

(4) Provides march security for moves between the reception site and the destina-

tion of the cargo or infiltrated personnel.

e. Recovery Party.

(1) Recovers cargo and air delivery equipment from the DZ.

(2) For air delivery operations, the recovery party should consist of at least 2 men for each parachutist or cargo container. The recovery party, usually dispersed along the length of the anticipated impact area, spots each parachute as it descends and moves to the landing point.

(3) Once a bundle is found, one man must stay with it while the second takes the parachute to the recovery collection point and guides a detail back to carry off the packages. Another technique is to divide the recovery party into 2-man teams which have been assigned a parachute number coinciding with the sequence of exit from the aircraft. If personnel are available, the recovery party leader stations a separate recovery detail at the far end of the DZ to track and locate bundles in event the exit is delayed or disrupted. Recovery party personnel must have a simple signaling means, such as a metal cricket or tone sticks, to preclude shouting and unnecessary movement. When the first bundle or parachutist exits from the aircraft, the recovery party leader should station

a man directly under the point of exit. This man remains in place until all bundles or parachutists are recovered. He serves as a reference for the point of exit and can subsequently indicate the aircraft's exact line of flight in the event a bundle is lost and a sweep of the DZ must be made.

(4) To insure sterilization, the recovery party must—

   (*a*) Clearly instruct all reception committee personnel on the dangers of leaving cigarette butts or candy and gum wrappers, mislaying equipment, and leaving "sign" of occupancy (crushed undergrowth, heel scuffs, trails, human waste).

   (*b*) When unpacking bundles, keep all rigging straps tied or buckled together; make only one cut on any single strap.

   (*c*) Have one individual at the recovery collection point to be responsible for accounting for air items and packages as the recovery teams bring them off the DZ.

   (*d*) Provide a two or three man surveillance team, preferably from the supporting auxiliary, to maintain a close watch on the DZ area for enemy activity during the 48 hours following the drop.

(5) To insure sterilization, the individual parachutist must—
- (*a*) Recover all parachute items; straps, buckles, pieces of equipment, or objects that have been introduced into the area in connection with the drop.
- (*b*) Not lay anything on the ground during the removal of individual parachutes (e.g., gloves, helmet, weapon) or during the recovery of bundles.
- (*c*) Bury unwanted air items separately at the base of thick bushes.
- (*d*) Erase drag marks, footprints, and impact marks with a scrubbing motion of a leafy tree branch.

    Note. Disguise freshly cut tree branch stubs with mud.
- (*e*) Avoid trampling or crushing vegetation; skirt around plowed areas and grass fields when moving off the DZ.
- (*f*) Select and prepare cache sites concealed from ground and air daylight observation.
- (*g*) Prevent accidental compromise of the operation by avoiding paths and roads and by moving cross-country to the assembly point.

f. Transport Party.
(1) Moves items received to distribution points or caches.
(2) Consists of part, or all, of the members comprising the command, markings, and recovery parties.

(3) Uses available means of transportation such as pack animals and wagons.

74. Drop Zone Authentication

a. Air to Ground. The aircraft is required to arrive over the DZ within a specified time usually extending from 2 minutes before scheduled drop time to 2 minutes after. The DZ markings are displayed according to the schedule. Arrival during this specific time period, approach on the designated track, and flying at designated altitude is indication to the reception committee that the aircraft is friendly.

b. Ground to Air.
  (1) The reception committee is identified by one of the following methods.
    (a) *Primary method.* Display a specific DZ marking configuration for the date or each day of the week. The display of proper markings for a particular 24-hour period identifies the reception committee.
    (b) *Alternate method.* Display a standard marking configuration indicating the release point and identifying the reception committee by means of a coded light or panel signal. The following rules govern the use of a coded light signal:
      1. Never use code letters consisting solely of all dots or dashes, e.g., I,E,-M,O,S,T.

  2. Use the following time intervals to assist the aircrew in recognition of the signal: dots—2 seconds; dashes—4 seconds; interval between dots and dashes—2 seconds; interval between coded letters—5 seconds.

 (2) The schedule of DZ markings, or identification and authentication signals, is contained in the detachment SOI. This schedule is changed as frequently as necessary for security.

Section VI. AIR LANDING OPERATIONS

75. General

Air landing operations provide the Special Forces detachment with a speedy and efficient means of evacuating both personnel and cargo from the operational area; however, such operations are difficult and require highly trained aircrews and reception committees. Normally, in guerrilla operational areas, air landing operations are conducted at night. In a counterinsurgency environment or in the conduct of counterguerrilla operations, air landing operations may be conducted either at night or in daylight commensurate with the mission.

76. Security

Special emphasis is placed on security. Because of the probability of the enemy's detecting the operation, LZ's established in guerrilla operational areas are located away from the guerrilla bases and are seldom reused. LZ's used in counterinsurgency operations may be adjacent to base camps rather than at great distances. This permits the counterinsurgency force to provide maximum security to POL dumps and areas used for loading and unloading equipment and for personnel moving in or out of the operational area.

77. Landing Zone (Land)

a. General. The general considerations applicable to DZ selection apply to the selection of LZ's; however, site size, approach features, and security are far more important.

b. Selection Criteria.

(1) *Terrain.*

(a) LZ's should be located in flat or rolling terrain.

(b) Level plateaus of sufficient size can be used. Because of decreased air density, landings at higher elevations require increased minimum LZ dimensions. If the LZ is located in terrain above 1,220 meters or in an area with a very high temperature, the minimum length will be increased as follows:

1. Add 10 percent to minimums for each 305 meters of altitude above 220 meters.

2. Add 10 percent to minimums for the

Figure 15. Landing zone (land) medium aircraft (night operations).

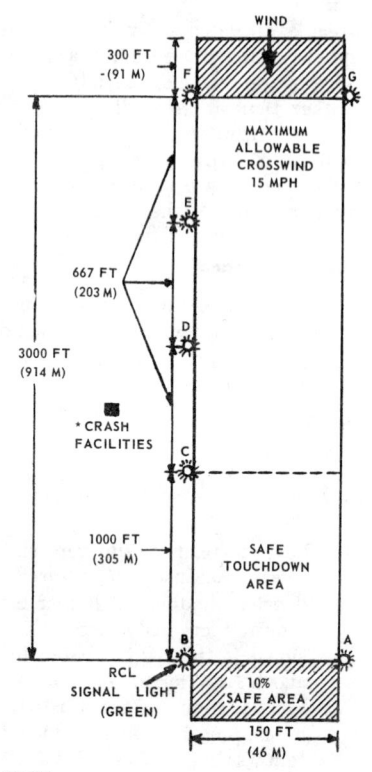

Figure 16. Landing zone (land) light aircraft (night operations).

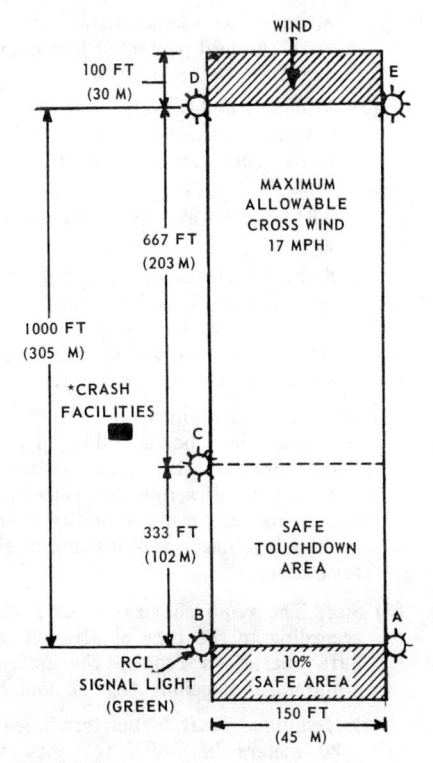

altitude if temperatures are over 90°F. Add 20 percent if temperatures are over 100°F (38°C).

- (c) Although undesirable, sites with only a single approach can be used. When using such sites it is mandatory that—
  1. All takeoffs and landings be made upwind.
  2. Sufficient clearance at either end of the LZ permits a level 180° turn to either side within a radius of 5 kilometers for medium aircraft, or 1.5 kilometers for light aircraft.

- (2) *Weather*. Prevailing weather in the landing area should be favorable. In particular, there must be a prior determination of wind direction and velocity and conditions restricting visibility such as ground fog, haze, or low-hanging cloud formations.

- (3) *Size*. The required size of LZ's varies according to the type of aircraft used. Safe operations require the following minimum dimensions (fig. 15 and 16):
  - (a) Medium aircraft, 920 meters in length, 30 meters in width (45 meters at night).
  - (b) Light aircraft, 305 meters in length, 15 meters in width (45 meters at night).

(c) In addition to the basic runway dimensions, and to provide a safety factor, these extra clearances are required—

    1. A cleared surface, capable of supporting the aircraft extending from each end of the runway and equal to 10 percent of the runway length.

    2. A 15 meter strip extending along both sides of the runway and cleared to within 1 meter of the ground.

(4) *Surface.*

  (a) The surface of the LZ must be level and free of obstructions such as ditches, deep ruts, logs, fences, hedges, low shrubbery, rocks larger than a man's fist, or grass over .5 meters high.

  (b) The subsoil must be firm to a depth .6 meters.

  (c) A surface containing gravel and small stones or thin layers of loose sand over a firm layer of subsoil is acceptable. Plowed fields or fields containing crops over .5 meters high will not be used.

  (d) Surfaces that are not desirable in summer may be ideal in winter. Ice with a thickness of 20 centimeters will support a light aircraft. Ice with a thickness of 61 centimeters will support a medium aircraft. Unless the aircraft is equipped for snow landing, snow in excess of 11 centimeters must be packed or removed from the landing strip.

(e) The surface gradient of the LZ should not exceed 2 percent.

(5) *Approach and takeoff clearance.* The approach and takeoff clearances are based on the glide/climb characteristics of the aircraft. For medium aircraft the glide/climb ratio is 1 to 40; that is, one meter of gain or loss of altitude for every 40 meters of horizontal distance traveled. The ratio for light aircraft is 1 to 20. As a further precaution, any obstructions in approach and departure lanes must conform to the following specifications (fig. 17):

(a) No obstruction higher than 2 meters and closer than 36 meters for light aircraft or closer than 72 meters for medium aircraft.

(b) A 15 meter obstruction may not be closer than 610 meters for medium aircraft, or 305 meters for light aircraft.

(c) A 155 meter obstruction may not be closer than 7 kilometers for medium aircraft or 4 kilometers for light aircraft.

(d) Hills of 305 meters or more above LZ altitude may not be closer than 13 kilometers from the landing zone for medium aircraft.

(e) The heights of obstacles are computed from the level of the landing strip.

(f) Distances are computed from the nearest end of the landing surface.

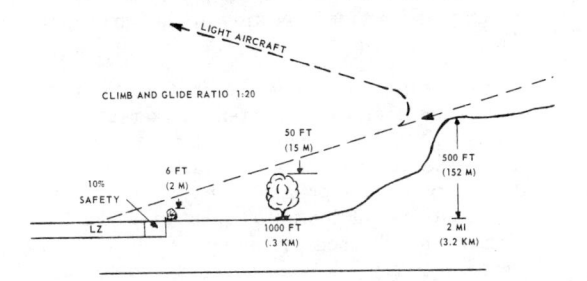

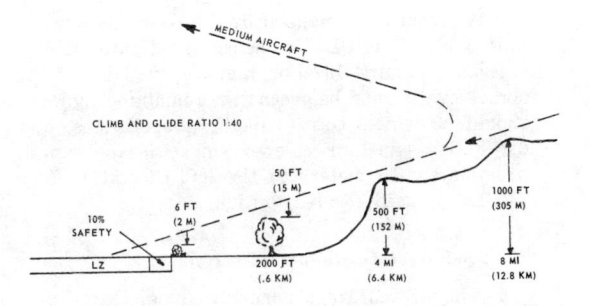

Figure 17. Takeoff and approach clearances (fixed-wing aircraft).

78. Markings

a. For night operations, lights are used for marking LZ's; during daylight panels are used. When flashlights are used, they should be handheld for directional control and guidance, and during the aircraft's final approach they should be held at knee height to avoid giving an erroneous impression of the location of the surface of the landing strip.

b. The pattern outlining the limits of the runway consists of five or seven marker stations (fig. 18). Stations A and B mark the downwind end of the LZ and are positioned to provide for the safety factors previously mentioned. These stations represent the initial point at which the aircraft should touch the ground. Station C indicates the very last point at which the aircraft can touchdown and complete a safe landing.

c. A signal light manned by the reception committee leader (RCL) is incorporated into light station B, figure 18. For night operations, the signal light should be green; the remaining lights should be white. During daylight operations, a distinctive panel or colored smoke located approximately 15 meters to the left of station B (RCL), is used for recognition.

79. Conduct of Operations

a. The LZ markings normally are displayed 2 minutes before the arrival time indicated in the mission confirmation message. The markings remain displayed for a period of 4 minutes or until the aircraft completes its landing roll after touchdown (fig. 18).

 b. Identification is accomplished by—

  (1) The aircraft arriving at the proper time on a prearranged track.

  (2) The RCL flashing or displaying the proper code signal.

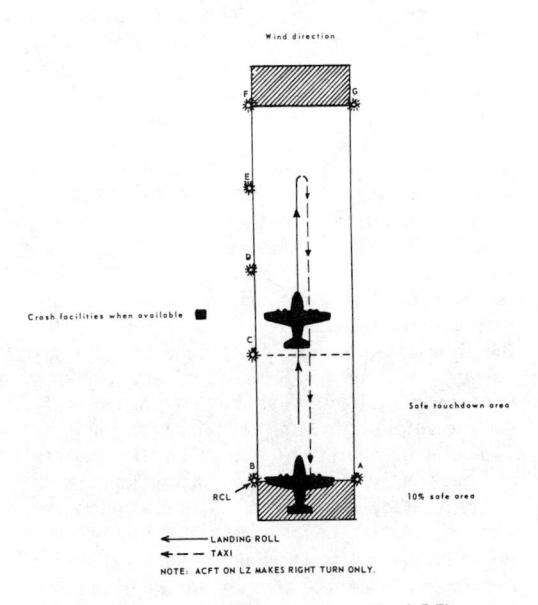

Figure 18. Landing procedure (land LZ).

c. Landing direction is indicated by—

(1) The RCL signal control light (station B) and marker A which are always on the approach or downwind end of the runway.

(2) The row of markers which are always on the left side of the landing aircraft.

d. The pilot usually attempts to land straight-in on the initial approach. When this is not possible, a modified landing pattern is flown using a minimum of altitude for security reasons. Two minutes before landing the RCL causes all lights of the LZ pattern to be turned on and aimed like a pistol in the direction of the aircraft's approach track. The RCL (station B) also flashes the code of the day continuously with the green control light in the direction of the aircraft's approach. Upon arrival in the area (within 15° to either side of the approach track and below 500 feet), the LZ marking personnel follow the aircraft with all lights. When the RCL determines that the aircraft is on its final approach he will cease flashing the code of the day and aim a steady light in the direction of the landing aircraft, being careful not to blind the pilot with the light. The solid light provides a more positive pattern and perspective for the pilot during landing. If a "go around" is required, all lights follow the aircraft to assist the pilot in maintaining orientation with respect to the landing strip; all lights continue to follow the aircraft during touchdown and until it passes each respective light station.

e. Landings normally are not made under the following conditions:

(1) Lack of, or improper, identification received from the LZ.

(2) When the RCL gives the abort signal, which is extinguishing of all lights on the LZ.

(3) Any existing condition that, in the opinion of the pilot, makes it unsafe to land.

f. After the aircraft passes the RCL position at touchdown and completes its landing roll and a right turn, stations A and B shine a solid light in the direction of the taxiing aircraft, continually exercising caution not to blind the pilot. These lights guide the pilot who will taxi the aircraft back to takeoff position. After off-loading or on-landing is complete and the aircraft is ready for takeoff, the RCL moves to a vantage point forward and to the left of the pilot, causes the LZ lights to be illuminated, and directs his light toward the nose of the aircraft as the signal for takeoff.

g. To eliminate confusion and insure expeditious handling, personnel or cargo to be evacuated or exfiltrated wait for unloading of incoming personnel or cargo.

h. When all evacuating personnel are loaded and members of the reception committee are clear of the aircraft, the pilot is given a go signal by the RCL. LZ markings are removed as soon as the aircraft is airborne.

i. Under ideal conditions, to increase the probability of success of an air landing operation, an alternate LZ may be designated for some missions. Separate landing times are established and both LZ's are manned. Personnel or cargo to be exfiltrated normally are stationed at the primary LZ; operational planning should allow for the orderly, secure transfer of personnel or cargo to the alternate LZ should the primary become unsuitable sufficiently in advance of the landing time. No attempt should be made to transfer personnel or cargo to the alternate LZ during the interval between landing times. Incoming personnel or cargo can be received at either the primary or alternate LZ.

Section VII. LANDING ZONE (WATER)

80. Selection Criteria

a. Size. For medium amphibious or seaplane aircraft, the required length is 1,220 meters with a minimum width of 460 meters; for light aircraft 615 meters long and 155 meters wide. An additional safe area equal to 10 percent of the airstrip length is required on each end (fig. 19).

b. Surface. Minimum water depth is 2 meters. The entire landing zone must be free of obstructions such as boulders, rock ledges, shoals, waterlogged boats, or sunken pilings within 2 meters of the surface.

c. Wind.

(1) Wind velocity for medium aircraft must

*WHEN AVAILABLE

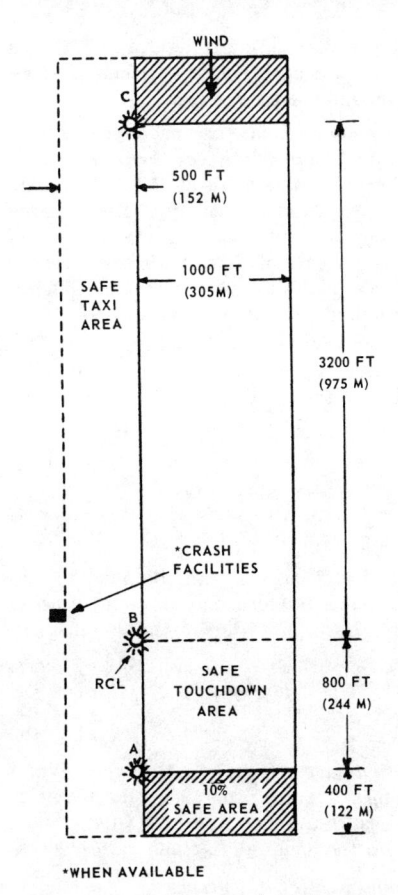

Figure 19. Landing zone (water) medium aircraft.

not exceed 20 knots, light aircraft, must not exceed 15 knots in either sheltered or semi-sheltered water.

(2) No landing can be made when crosswind components are greater than 8 knots from an angle of 45° to 90° to aircraft's landing heading. Medium and light aircraft can land in cross-wind components of 20 and 15 knots respectively at 0° to 40° angles from landing heading.

(3) Windwaves for medium aircraft will not exceed 1.9 meters in height, and windwaves for light aircraft will not exceed .5 meters in height.

d. Tide. Tides should have no bearing on the suitability of the landing area; however, whether high tide or low tide, the area should conform to minimum requirement.

e. Water/Air Temperatures. Because of the danger of icing, water and air temperatures must not fall below the following minimums:

	Water temperature	*Air temperature*
Salt Water	+18°F (—8°C)	+26°F (—3°C)
Fresh water	+35°F (+2°C)	+35°F (—2°C)
Brackish water	+30°F (—1°C)	+35°F (+2°C)

f. Approach and Takeoff Clearance. Water landing zones require approach/takeoff clearances identical to those of land LZ's (fig. 16) and are based on the same glide/climb ratios.

g. Marking and Identification.

(1) Depending upon visibility, lights or panels are used to mark water LZ's.

(2) The normal method of marking water LZ's is to aline three marker stations along the left edge of the landing strip

VELOCITY (KNOTS)	HEIGHT OF SEA (FEET)	SURFACE CONDITIONS
0	0	Smooth slick sea.
2	0	Small occasional ripples.
3-4	1/2	Small ripples all over - no calm areas.
5-6	1	Well defined wave - smooth with no breaking.
7-9	2	Occasional whitecaps.
10-11	3	Pronounced waves, frequent whitecaps which carry a short distance.
12-13	4	Whitecaps close together carrying over a distance equal to wave height; slight traces of wind streaks.
14-16	5	Clearly defined wind streaks whose lengths are becoming equal to about ten (10) wave lengths; light flurry patches.
17-19	7-8	Long well-defined streaks, coming from same direction as wind; many whitecaps.
20-22	9-10	Streaks are long and straight; whitecaps on every crest; wind picks up and carries mist along; large waves.

Figure 20. Wind and sea prediction table.

(fig. 19). Station A is positioned at the downwind end of the strip and indicates the desired touchdown point. Station B marks the last point at which the aircraft can touchdown and complete a safe landing. Station B is also the location of the RCL and the pickup point. The RCL light is an additional light and should not be in the same boat as the B station marker light. Station C marks the upwind extreme of the landing area. At night, stations A, B, and C are marked by white lights. The RCL signal light is green.

(3) An alternate method is to use a single marker station, marked at night with a steady light in addition to the signal or recognition light. This station is located to allow a clear approach and takeoff in any direction. The pilot is responsible for selecting the landing track and may touchdown on any track 305 meters from the marker station. Following pickup, the aircraft taxis back to the 610 meter circle in preparation for takeoff (fig. 21).

h. Conduct of Operations.

(1) The LZ is carefully cleared of all floating debris and the marker stations are properly alined and anchored to prevent drifting.

(2) The procedure for displaying water LZ

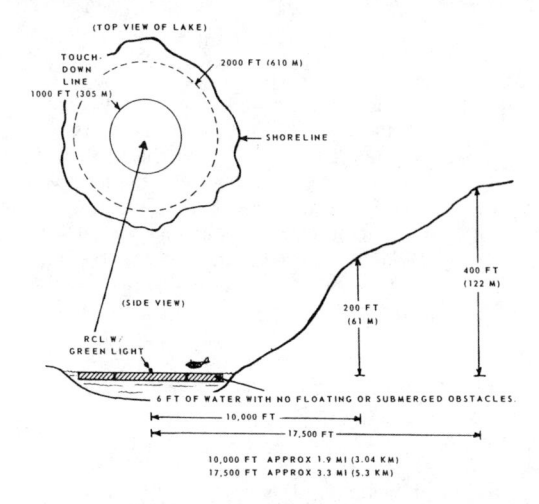

Figure 21. One light water landing zone (night).

markings and identification is the same as for operations on LZ's.

(3) Personnel or cargo to be evacuated or exfiltrated are postioned in the RCL boat. Following the landing run, the aircraft turns to the left and taxis back to the vicinity of the RCL boat to make the pickup. The RCL indicates his position by shining the signal light toward the aircraft and continues to shine his light until the pickup is completed. This light should not be aimed directly at the

cockpit to avoid blinding the aircrew (fig. 22).

(4) The RCL boat remains stationary during pickup operations. The aircraft taxis to within 15 to 30 meters of the RCL boat, paying out a dragline from the left rear door. The dragline is approximately 45 meters in length and has three life jackets attached; one close to the aircraft, a second at the midpoint, and the third on the extreme end of the line. The life jackets have small marker lights attached during night operations. The aircraft taxis to the left around the RCL boat, bringing the dragline close enough to be secured. The RCL fastens the line to the boat (fig. 23). Due to the danger of swamping the craft, the RCL does not attempt to pull on the line. Members of the aircrew pull the boat to the door of the aircraft. Should the boat pass the aircraft door and continue toward the front of the aircraft, all personnel in the boat must abandon immediately to avoid being hit by the propeller (fig. 23).

(5) After pickup, the aircrew is given any information that will aid in the takeoff. Following this, the RCL boat moves to a safe distance from the aircraft and signals the pilot "all clear." If necessary, previously installed JATO bottles are used for positive power takeoff.

←——— LANDING
←--- TAXI

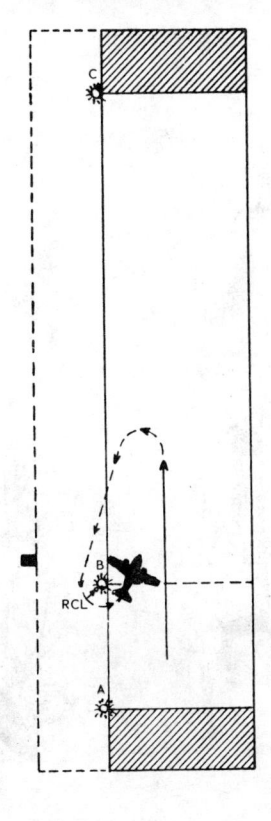

Figure 22. Landing procedures (water LZ).

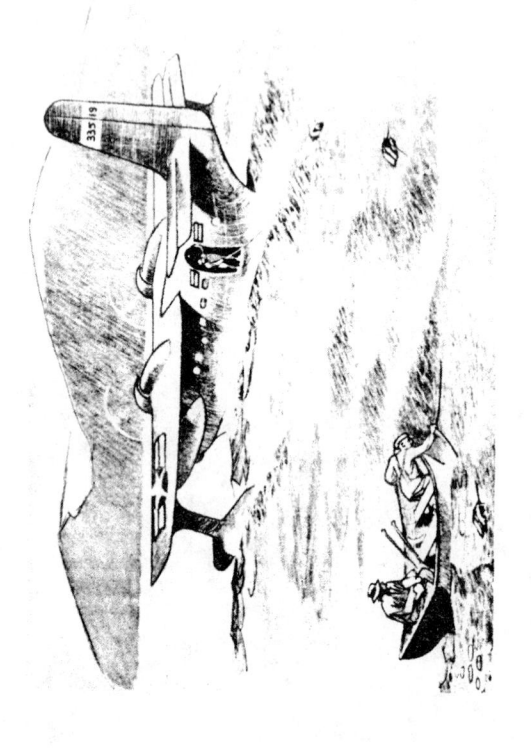

Figure 23. Water pick-up operations.

(4) Heavy dust or loose snow interferes with pilot visions just before touchdown. This effect can be reduced by clearing, wetting down, or using improvised mats.

(5) Landing pads may be prepared on swamp or marsh areas by building platforms of locally available materials (fig. 25). Such LZ's are normally used for daylight operations only. The size of the clearing for this type of LZ is the same

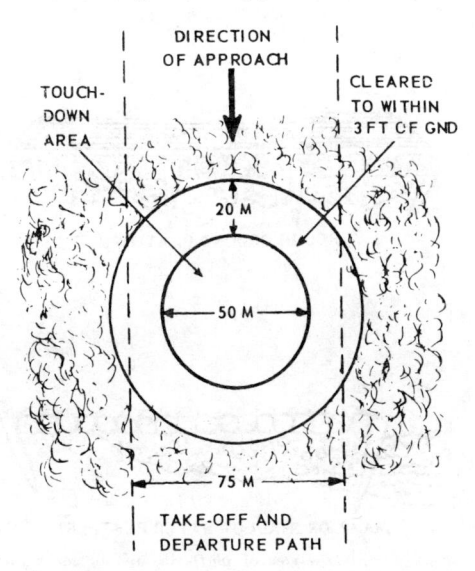

Figure 24. Landing zone for rotary-wing aircraft.

as *b* above, with the following additional requirements for the platform:

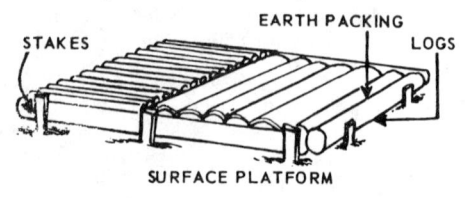

Figure 25. Examples of platform landing zones for rotary-wing aircraft.

(a) Large enough to accommodate the the spread of the landing gear (plus 10 feet).
   (b) Capable of supporting the weight of the aircraft.
   (c) Of firm construction that will not move when the helicopter touches down and rolls slightly forward.
   (d) Level.
   (e) If logs or bamboo are used, be constructed so that the top layer of poles is at right angles to the touchdown direction.

  (6) Helicopters, other than the HU-1 series, can land in water without the use of special flotation equipment provided—
   (a) The water depth does not exceed 46 cm.
   (b) A firm bottom such as gravel or sand exists.

  (7) Landing pads can be prepared on mountains or hillsides by cutting and filling (fig. 26). Caution must be exercised to insure there is adequate clearance for the rotors.

c. *Approach/Takeoff.*

  (1) There should be at least one path of approach to the LZ measuring 75 meters in width.
  (2) A rotary-wing aircraft is considered to have a climb ratio of 1 to 5 (fig. 27).
  (3) Takeoff and departure from the LZ may be along the same path used for the ap-

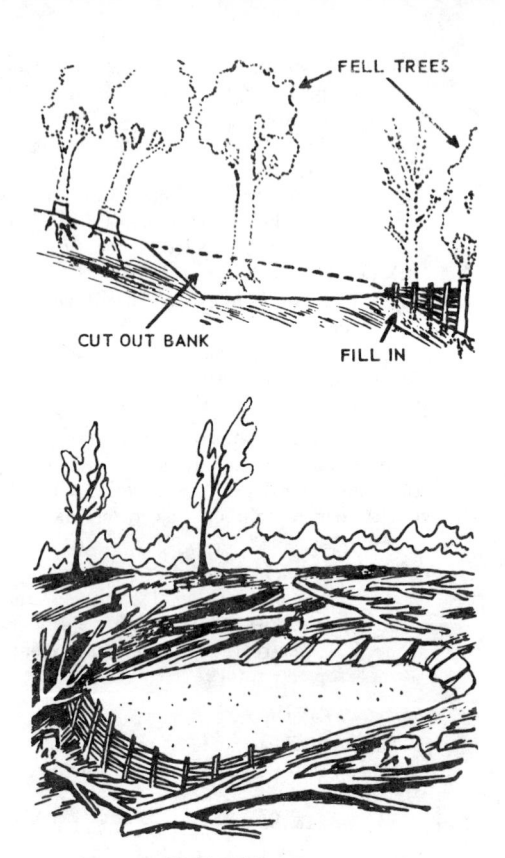

Figure 26. Preparing landing pads in mountainous terrain (rotary-wing aircraft).

proach; however, a separate departure path as free from obstacles as the approach path is desired (fig. 27).

83. Marking and Identification

a. LZ's for rotary-wing aircraft are marked to—

(1) Provide identification of the reception committee.
(2) Indicate direction of wind or required direction of approach.
(3) Delineate the touchdown area.

b. Equipment and techniques of marking are similar to those used with fixed-wing LZ's—lights or flares at night and panels in daylight.

c. An acceptable method of marking is the Y system. This uses four marker stations (fig. 28).

(1) The direction of approach is into the open end of the Y.

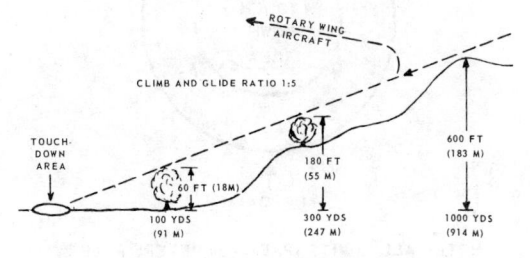

Figure 27. Approach/takeoff clearances (rotary-wing aircraft).

(2) When compatible with approach paths, wind direction is along the stem of the Y toward the open end.

(3) The touchdown area is delineated by the triangle formed by the three lights marking the open end of the Y.

(4) Station No. 2 is also the signal station. Light or panel signals may be used for identification. Smoke may be used to

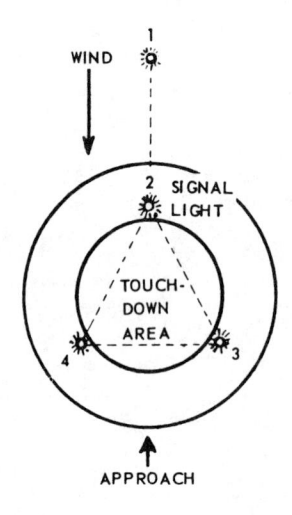

NOTE: ALL LIGHTS SPACED 50 METERS APART

Figure 28. Markings of landing zones for use by rotary-wing aircraft.

assist the pilot in locating the landing zone, but it is placed so it will not obscure the touchdown area.

84. Reporting Landing Zones

a. Reporting of LZ's and the coordination between the operational detachment and the air support unit through the SFOB closely parallels the procedures used in aerial delivery operations. The minimum LZ data required is—

   (1) *Code name.* Extracted from detachment SOI.
   (2) *Location.* Complete military grid coordinates of the center of the LZ.
   (3) *Long axis.* Magnetic azimuth of long axis of runway. It also indicates probable direction of landing approach based on prevailing winds.
   (4) *Description.* Surface, length, and width of runway.
   (5) *Open Quadrant.* Measured from center LZ and reported as series of magnetic azimuths. Open quadrants indicate acceptable aircraft approach.
   (6) *Track.* Series of magnetic azimuths clockwise from north.
   (7) *Obstacles.* In addition to the limitations pertaining to takeoff and approach clearances, obstacles exceeding 90 meters above the level of the LZ within a 8 kilometer radius, and not shown on issued maps, are reported by description, azi-

muth, and distance from the center of the LZ.

(8) *Reference point.* A landmark shown on the issued map; reported by name, azimuth, and distance from center of LZ. Used with (2) above in plotting the LZ location.

(9) *Date/time mission requested.*

(10) *Items to be infiltrated or evacuated.* See appendix III for a sample LZ report.

b. Reporting LZ's for Rotary-Wing Aircraft. The minimum landing zone data reported generally is the same as for fixed-wing LZ's, except in first paragraph indicate that the site is for use by rotary-wing aircraft.

85. Mission Confirmation (Air Landing)

a. Following the processing of the LZ data at the SFOB, and coordination with the air support unit, a confirmation message is transmitted to the operational detachment. This procedure is similar to that used for aerial delivery operations.

b. The confirmation message contains, as a minimum, the code name of the LZ, the date/time that the aircraft will arrive, and the track to be flown.

c. See appendix III for sample air landing confirmation message.

86. Skyhook Operations

a. General. Exfiltration is the means employed

to return or bring personnel out of operational areas to friendly territory. Exfiltration and evacuation techniques normally employed in support of unconventional warfare and counterinsurgency operations have already been discussed in the air landing portion of this chapter. Skyhook is another technique within the capabilities of the Special Forces groups, Army aviation support units, and Air Force support units. Skyhook techniques may be used to exfiltrate or evacuate personnel and equipment from areas inaccessible to aircraft landings.

b. Missions. Skyhook missions must be approved by the SFOB and the supporting air unit selected for the mission. Aircraft equipped to fly exfiltration or evacuation missions using skyhook techniques will be placed on standby status. Operational missions may include evacuation or infiltration of—

(1) Seriously ill or injured U.S. personnel.
(2) Guides or assets who can brief operational elements and reinfiltrate with operational detachments.
(3) Priority and valuable cargo and equipment from remote areas, that might normally require days or weeks of hazardous travel to bring out.
(4) Downed aircrews.
(5) Bodies subject to possible desecration such as heroes and martyrs.
(6) Personnel engaged in underwater operations against selected targets following mission accomplishment.

(7) Prisoners who possess useful information.

c. Capabilities. Because of the unique vertical lift capability of skyhook, combined with the use of large, long-range aircraft, geographical restrictions are few. Pickups can be performed under varying conditions and may include—

(1) Pickups performed in 40 knot winds.

(2) Night pickups with stroboscopic lights attached to the lift line.

(3) Open sea pickups.

(4) The use of small clearings in dense forests that will not permit helicopter landings.

(5) Multiple pickups, using longer and heavier duty lift lines. Loads are strung along the line separately rather than being tied together in one unit.

(6) A virtual all-weather capability. Weather affects operation only to extent that visibility may be restricted. Operations normally are delayed only when weather grounds aircraft.

d. Equipment.

(1) Skyhook equipment consists of 2 airdroppable containers of heavy duck material and nylon webbing. In the containers are two fiberglas containers filled with 650 cubic feet of helium; a polyethelene, dirigible-shaped balloon with an automatic valve that seals when

inflation is complete; 500 feet of tubular nylon lift line; a protective helmet; an all-weather, nylon coverall suit with zippered front, one chest strap, integral self-adjusting harness to fit any size person, and sheepskin hood protecting the head and neck. The equipment may be packaged in a waterproof container and equipped with a rubber life raft for airdropping into water or swampy areas. Difficulties in reading printed instructions are obviated by an animated cartoon instruction board.

(2) Aircraft (fig. 29) presently used by light aviation units Air Force support units, with slight modifications, may be used for skyhook operations. The aircraft is equipped with a "yoke" or wide fork horizontally mounted on its nose. The yoke is used only to guide the lift line to its center where it locks to the nose. The yoke with a 25-foot opening, is constructed of light weight tubing and readily supports a 200 pound weight. Deflection lines divert the lift line around the wing tips in the event the pilot fails to intercept the yoke. An electrical, hydraulic, or pneumatic winch is mounted inside the aircraft. Large radomes and side access hatches cause no interference.

87. Employment

a. Normal air drop techniques will be used to

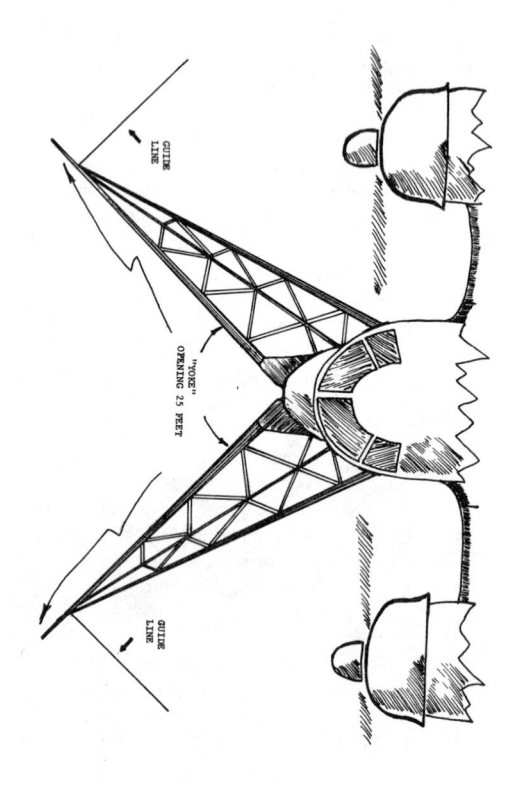

Figure 29. Top view of yoke mounted on aircraft.

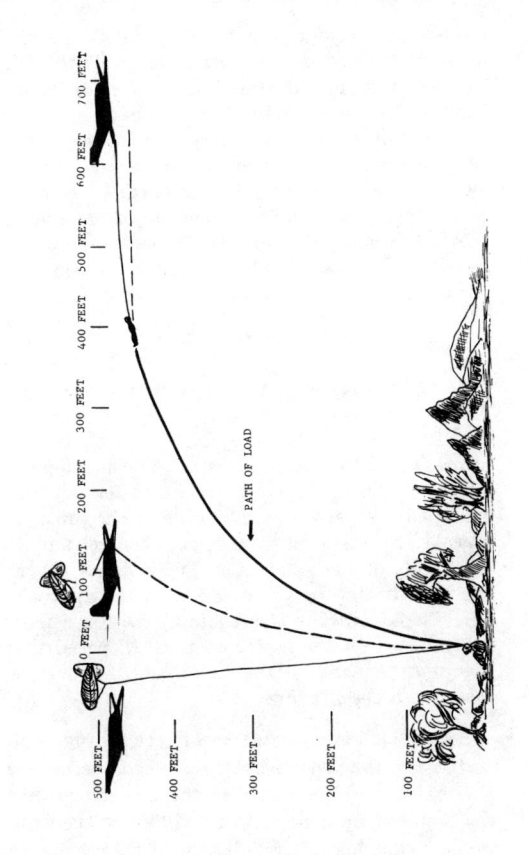

Figure 30. Skyhook pickup.

deliver the special equipment required. The equipment is dropped on the initial pass over the DZ. The person to be exfiltrated or evacuated, following directions on the illustration board, will don the coverall suit with the harness, attached to the lift line which is attached to the balloon. The balloon is plugged into the helium bottles. The exfiltrator then pulls a safety pin and squeezes a valve on the helium bottle, permitting the flow of gas to inflate the balloon. He releases the balloon, and sits down facing the approaching aircraft (fig. 30).

b. On the return pass over the DZ, the aircraft, flying upwind of the lift line, approaches the balloon at an altitude of approximately 400 feet. Three cerise colored flags, spaced 25 feet apart, are attached to the lift line; the first flag is 50 feet below the balloon and serves as the contact point for the aircraft. In night operations the lift line, equipped with stroboscopic lights at the same intervals are activated by a remote control unit in the hands of the exfiltrator. These lights permit the pilot to line up the aircraft on the approach run. The aircraft is also equipped with a remote control unit on the instrument panel, permitting the pilot to activate the lights until contact is made with the lift line.

c. Special Forces detachments requesting exfiltration or evacuation by the skyhook technique will be required to establish the DZ and receive the required equipment. They will assist the exfiltrator to don his suit and dispose of the remaining equipment immediately after the pickup. In emergencies, the person to be exfiltrated may be required to perform the ground phase of the operation alone. In this event, areas selected for the pick-up should be remote and inaccessible to ground interference. Remaining equipment may be hidden by the person to be exfiltrated, or picked up at a later date by the operational detachment.

88. Low Level Parachute Extraction Resupply System (LOLEX)

a. General. This system of resupply is the delivery of supplies by cargo aircraft without the aircraft's landing in the operational areas. This system is within the capabilities of light aviation. Using the LOLEX technique, aircraft require the same landing patterns, turn-around radio, and approach and takeoff clearances as normal landing operations.

b. Extraction. Extraction of loads up to 6,000 pounds, depending on the aircraft, is accomplished while the aircraft flies above ground at delivery point, altitude 3 to 6 feet. The extraction parachute deploys and pulls the load from the rear of the aircraft. The deceleration provided by the parachute's opening and ground friction quickly stops the forward momentum of the load.

c. Lashing. The load normally is lashed to wooden platforms with the forward portion rounded like the front ends of skis or sleds. This prevents the load's digging into the ground and flipping over.

d. Advantages. The advantages of LOLEX are

pinpoint accuracy, ability of the air support unit and reception committee to clear the landing zone quickly, and the absence of a requirement for ground equipment.

e. Responsibility. The reception committee must provide transport personnel sufficient to remove dropped loads as quickly as possible.

f. Further Information. Complete details on LOLEX operations are found in TM 57–210–1.

CHAPTER 7
RAIDS AND AMBUSHES

Section I. RAIDS

89. Raid Techniques

The raid is one of the basic operational techniques employed by Special Forces in both unconventional warfare and counterinsurgency operations. The keywords to the successful accomplishment of any raid are flexibility and responsiveness to orders and direction. In preparation for raids in counterinsurgency operations, Special Forces detachment commanders should have plans (ready for implementation if necessary) for the use of artillery support, close air and aerial fire support, reinforcements, and immediate evacuation.

a. General. A raid is a surprise attack against an enemy force or installation. Such attacks are characterized by secret movement to the objective area; brief, violent combat; rapid disengagement from action; and swift, deceptive withdrawal. Raids are conducted to destroy or damage supplies, equipment, or installations such as command posts, communication facilities, depots, or radar sites; to capture supplies equipment and key personnel; or to cause casualties among the enemy and his supporters. Raids also serve to distract attention from other operations keep the enemy off balance, and force deployment of additional units to protect rear areas.

b. Organization of the Raid Force.

(1) *Size.* The size of the raid force depends upon the mission, nature and location of the target, and the enemy situation. The raid force may vary from a squad attacking a checkpoint or a portion of unprotected railroad track to a battalion attacking a large supply depot. Regardless of size, the raid force consists of two basic elements—assault and security.

(2) *Assault element.* The assault element is organized and trained to accomplish the objectives of the raid. It consists of a main action group to execute the raid mission and may include personnel detailed to execute special tasks which aid the main action group.

(*a*) The main action group executes the major task insuring the success of the raid. For instance, if the raid objective is to destroy a critical installation such as a railroad bridge or tunnel, the main action group emplaces and detonates the demolition charges. If the target, such as enemy personnel, is to be neutralized by fire, the main action group conducts its attack with a high proportion of automatic weapons. In some instances the main action group moves physically on or into the target; in other instances they are able to accomplish their task from a distance. The other elements of the raid force are designed to allow the main action group access to the target for the time required to accomplish the raid mission.

    (b) If required, special task details assist the main action group to reach the target. They execute such complementary tasks as eliminating guards, breaching and removing obstacles to the objective, diversionary or holding actions, and fire support. The special task details may precede, follow, or act concurrently with the main action group.

  (3) *Security element.* The security element supports the raid by preventing the enemy's reinforcing or escaping the target area. The security element also covers the withdrawal of the assault element and acts as a rear guard for the raid force. The size of the security element depends upon the enemy's capability to intervene and disrupt the operation.

c. Preparation.

  (1) *Planning considerations.*

(a) The first step is the selection of the target, based upon criticality, vulnerability, accessibility, and recuperability. Other important considerations are the nature of the terrain and the combat efficiency of the raiding force.

(b) Secondly, the Special Forces and indigenous force commanders must consider any possible adverse effects on their units and the civilian populace as a result of the raid. The objective is to diminish the enemy's military potential, but an improperly timed operation may provoke enemy counteraction for which indigenous units and the populace are unprepared. Also, an unsuccessful attack often has disastrous effects on troop morale while successful operations, on the other hand, raise morale and increase the prestige of the units and their leaders in the eyes of the civilians and makes them more willing to provide much needed support. Further, every precaution is taken to insure that civilians are not needlessly subjected to harsh reprisals because of raid actions. The impact of successful raids can be exploited in detachment psychological operations; however, it is important that before such action is taken, any possible unfavorable repercussions from the population and the enemy military forces be considered. If a raid is unsuccessful, psychological operations will be required to lessen any adverse effects on the friendly indigenous force.

(c) Although detailed, the plan for a raid must be simple and not depend upon too many contingencies for success. Activities in the objective area are planned so that the installation to be attacked is not alerted. This means that activities will conform to normal patterns. Time and space factors are carefully considered—time is allowed for assembly and movement, particularly during darkness. All factors are considered to determine whether movement and attack should be made during daylight or darkness. Darkness naturally favor surprise and normally is the best time when the operation is simple and the physical arrangement of the installation well known. Early dawn or dusk is favored when inadequate knowledge of the installation or other factors necessitate tight control of the operation. A withdrawal late in the day or at night makes close pursuit by the enemy more difficult.

(2) *Intelligence.* The raid force commander must have maximum intelligence of the target, enemy forces capable of inter-

vening the civilian population's attitude and support, and the terrain to be traversed en route to and from the objective area; therefore, an intensive intelligence effort precedes the raid. Indigenous intelligence and reconnaissance elements conduct premission reconnaissance of the route to the target and of the target itself. In guerrilla operations local auxiliary sources are exploited, and the auxiliaries may act as guides. Surveillance of the target begins early and is continuous up to the time of the attack. The raid force commander exercises extreme caution to insure the secrecy of the impending operation by careful assignment of missions to indigenous reconnaissance elements so that the local population will not become alerted and alarmed.

(3) *Rehearsals of participants.* Realistic rehearsals by all participants are conducted for the operation; terrain similar to that found in the target area is used when available; sand tables, sketches, photographs, and target mockup are used to assist in briefings; contingency actions are practiced, and final rehearsals are conducted under conditions of visibility expected in the objective area.

(4) *Final inspection.* The raid force commander conducts a final inspection of

personnel and equipment before moving to the objective area. If possible, weapons are test fired, faulty equipment is replaced, and the condition of the men is checked. During this inspection a counterintelligence check is made of personal belongings to insure that no incriminating documents are carried during the operation. This inspection assures the raid force commander that his unit is equipped and ready for the operation.

d. Movement (fig. 31). Movement to the objective area is planned and conducted to allow the raid force to approach the target undetected. Movement may be over single or multiple routes. The preselected route or routes terminate in or near one or more mission support sites. Every effort is made to avoid contact with the enemy during movement. Upon reaching the designated rendezvous and mission support sites, security groups are deployed and final coordination takes place before moving to the attack position.

e. Action in the Objective Area (fig. 32). Special-task details move to their positions and eliminate sentries, breach or remove obstacles, and execute other assigned tasks. The main action group quickly follows the special-task details into the target area. Once the objective of the raid has been accomplished the main action of the group withdraws, covered by preselected fire support elements or part of the security force. If the attack is unsuccessful, the action is terminated to prevent undue loss and the special task details withdraw according to plan. The asault element assembles at one or more rallying points while the security elements remain in position to cover the withdraw according to plan. The assault element withdraw on signal or at a prearranged time.

f. Withdrawal (fig. 33).

(1) Withdrawal is designed to achieve maximum deception of the enemy and facilitate further action by the raid force. The various elements of the raid force withdraw, on order, over predetermined routes through a series of rallying points. Should the enemy organize a close pursuit of the assault element, the security element assists by fire and movement, distracting the enemy, and slowing him down. Elements of the raid force which are closely pursued by the enemy do not attempt to reach the initial rallying point; but, on their own initiative, they lead the enemy away from the remainder of the force and attempt to lose him by evasive action in difficult terrain. If the situation permits, an attempt is made either to reestablish contact with the raid force at other rallying points, to continue to the base area as a separate group, or to reach an area for evaluation. The raid force, or elements of it, may separate and proceed as small groups or individuals to evade close pursuit.

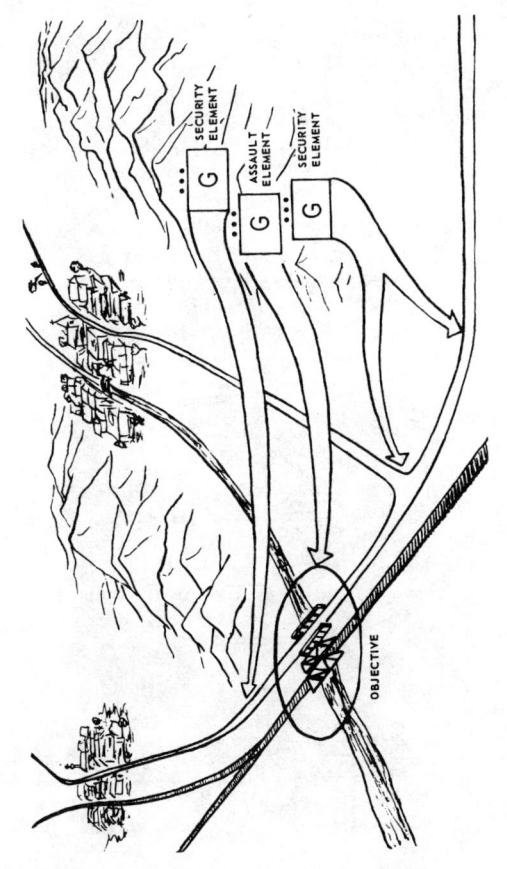

Figure 31. Raid-movement to the objective.

Figure 32. Raid action in the objective area.

(2) Frequently, in GWOA's the raid force disperses in smaller units, withdraws in different directions, and reassembles at a later time at a predesignated place to conduct further operations. Elements of the raid force can conduct other operations, such as an ambush of the pursuing enemy force, during the withdrawal.

g. Large Raids.

(1) *General.* When a target is large, important to the enemy, and well guarded, a larger raid force is required to insure an attack. Large raids may involve the use of a battalion-size unit; and, though the operation is conducted similarly to that for smaller raids, additional problems must be considered (fig. 34).

(2) *Movement to objective area.* Surprise is just as desirable as in a smaller raid, but it is usually harder to achieve. In GWOA's, the number of troops to be deployed requires additional mission support sites at a greater distance from the target to preserve secrecy, which necessitates a longer move to the attack position. A large-raid force usually moves by small components over multiple routes to the objective area. In counterinsurgency operations larger raids are conducted by regular troops, at times using air assault techniques,

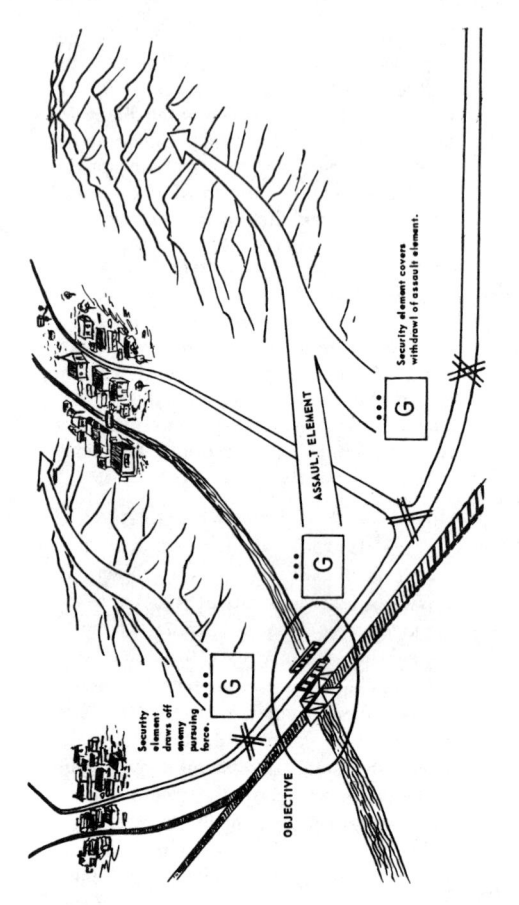

Figure 33. Raid withdrawal from action.

supported by Special Forces-directed units.

(3) *Control.* Another problem inherent in a large raid is that of control. Units without extensive radio communications equipment will find coordination of widespread elements difficult to achieve. Pyrotechnics, audible signals, runners, or predesignated times may be used to coordinate action.

(4) *Training.* A high degree of training and discipline is required to execute a large raid. Extensive rehearsals assist in preparing the force for the mission. In particular, commanders and staffs must learn to use large numbers of troops as a cohesive fighting force.

(5) *Fire support.* Additional fire support usually is a requirement. In GWOA's this may mean secretly caching ammunition in mission support sites over a period of time before the raid. Guerrillas may each carry a mortar, recoilless rifle round, rocket or box of machinegun ammunition and leave them at a mission support site or firing position for fire support units. In counterinsurgency operations sufficient firepower is moved into areas within range of the selected target area and is on call to support the action.

(6) *Timing.* Timing is usually more difficult for a large raid. More time is required

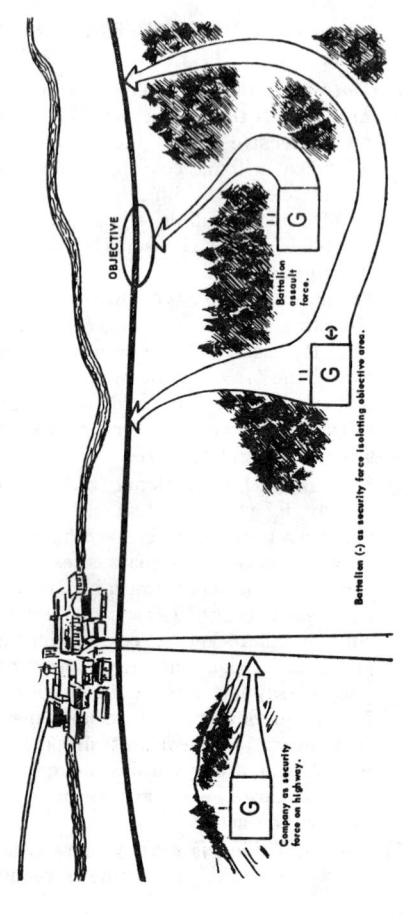

Figure 34. Large raid.

to move units, and the main action element needs more time to perform its mission. This requires stronger security elements to isolate the objective for longer periods. The timing of the raid takes on increased importance because of the large numbers of personnel involved. Movement to the objective is usually accomplished during periods of low visibility; however, because of fire support coordination requirements and larger numbers of personnel, the action may take place during daylight hours.

(7) *Withdrawal.* In a GWOA, withdrawal from a large raid usually is by smaller groups over multiple routes in order to deceive the enemy and dissipate his pursuit. Dispersed withdrawal has the added advantage of denying a lucrative target to enemy air and fire support elements; however, the raid force commander must consider the possibility of an alert and aggressive enemy defeating the dispersed elements of the force. All factors must be carefully weighed before deciding on how to conduct the withdrawal.

Section II. AMBUSHES

90. Techniques

a. General. An ambush is a surprise attack from a concealed position, used against moving or temporarily halted targets such as trains, boats, truck convoys, individual vehicles, and dismounted troops. In an ambush the enemy sets the time and the attacker sets the place. Ambushes are conducted to destroy or capture personnel and supplies, harass and demoralize the enemy, delay or block movement of personnel and supplies, and channel enemy movement by making certain routes useless for traffic. They usually result in concentrating the majority of movements to principal roads and railroads where targets are more vulnerable to attack by other forces.

 (1) *Organization.* Like the raid force, the ambush force is organized into assault and security elements. The assault element conducts the main attack against the ambush target which includes halting the column, killing or capturing personnel, recovering supplies and equipment, and destroying unwanted vehicles or supplies which cannot be moved. The security force isolates the ambush site using roadblocks, other ambushes, and outposts. Security elements cover the withdrawal of the assault element.

 (2) *Preparation.* Planning and preparing an ambush is similar to planning and preparing a raid except that selecting the ambush site is an additional consideration.

  (*a*) The mission may be a single ambush against one column or a series of ambushes against one or more routes of communication.

(b) The probable size, strength, and composition of the enemy force that is to be ambushed, formations likely to be used, and enemy reinforcement capabilities are considered.

(c) Favorable terrain for an ambush, providing unobserved routes for approach and withdrawal, must be selected.

(d) Time of the ambush should coincide with periods of low visibility, offering a wider choice of positions and better opportunities to surprise and confuse the enemy; however, movement and control are more difficult during the night ambush. Night ambushes are more suitable when the mission can be accomplished during, or immediately following, the initial burst of fire. They require a maximum number of automatic weapons to be used at close range. Night ambushes can hinder the enemy's use of routes of communications at night, while friendly aircraft can attack the same routes during the day. Daylight ambushes facilitate control and permit offensive action for a longer period of time and provide the opportunity for more effective fire from such weapons as rocket launchers and recoilless rifles.

(3) *Intelligence.* Since the guerrillas are sel-

dom able to ascertain the exact composition, strength, and time of convoy movements in advance, their intelligence effort should be directed towards determining the convoy pattern of the enemy. Using this information, guerrilla commanders are able to decide on the convoys to be attacked by ambush. Intelligence considerations for a raid are equally applicable to an ambush.

(4) *Site Selection.* In selecting the ambush site, the basic consideration is favorable terrain, although limitations such as deficiencies in firepower and lack of resupply during actions may govern the choice of the ambush site. The site should have firing positions offering concealment and favorable fields of fire. Whenever possible, firing should be through a screen or foliage. The terrain at the site should serve to funnel the enemy into a killing zone. The entire killing zone is covered by fire so that dead space that would allow the enemy to organize resistance is avoided. The ambush force should take advantage of natural obstacles such as defiles, swamps, and cliffs to restrict enemy maneuvers against the force. (When natural obstacles do not exist, mines, demolitions, camouflaged barbed wire, and other concealed obstacles are employed to canalize the enemy.) Security elements are placed

on roads and trails leading to the ambush site to warn the assault element of the enemy approach. These security elements also assist in covering the withdrawal of the assault element from the ambush site. The proximity of security to assault elements is dictated by terrain. In many instances, it may be necessary to organize secondary ambushes and roadblocks to intercept and delay enemy reinforcements.

b. Conducting the Ambush.

(1) *Movement.* The ambush force moves over a preselected route or routes to the ambush site. One or more mission support sites or rendezvous points usually are necessary along the route to the ambush site. Last minute intelligence is provided by reconnaissance elements, and final coordination for the ambush is made at the mission support site.

(2) *Action at the ambush site* (fig. 35).

(a) Troops are moved to an assembly area near the ambush site, and security elements take up their positions first and then the assault elements move into place. As the approaching enemy column is detected, or at a predesignated time, the ambush commander decides whether or not to execute the ambush. This decision depends upon the size of the enemy column, guard and security measures, and estimated worth of the target in light of the mission. If the decision is made to execute the ambush, advance guards are allowed to pass through the main position. When the head of the main column reaches a predetermined point, it is halted by fire, demolitions, or obstacles. At this signal the entire assault element opens fire. Designated details engage the advance and rear guards to prevent reinforcement of the main column. The volume of fire is rapid and directed at enemy personnel exiting from vehicles and concentrated on vehicles mounting automatic weapons. Antitank grenades, rocket launchers, and recoilless rifles are used against armored vehicles. Machineguns lay bands of fixed fire across escape routes. Mortar shells, hand and rifle grenades are fired into the killing zone. If the commander decides to assault, it is launched under covering fire on a prearranged signal. After enemy resistance has been nullified, special parties move into the column to recover supplies, equipment, and ammunition. When the commander desires to terminate the action, because the mission either has been accomplished or superior enemy reinforcements are arriving, he withdraws the assault element first and then the security elements which cover the withdrawal of the assault element.

(b) If the purpose of the ambush is to harass and demoralize the enemy, a different tactic may be adopted. The advance guard is selected as the target of the ambush and the fire of the assault element is directed against them. Repeated attacks against enemy advance guards—

1. Cause the use of disproportionately strong forces in advance guard duties. This may leave other portions of the column vulnerable or require the diversion of additional troops to convoy duty.

2. Create an adverse psychological effect upon enemy troops, and the continued casualties suffered by the advance guard make such duty unpopular.

(3) *Withdrawal.* Withdrawal from the ambush site is similar to withdrawal from a raid in that the security elements cover the assault elements.

c. *Special Ambush Situations.*

(1) *Columns protected by armor.* Attacks against columns protected by armored vehicles depend upon the type and location of armored vehicles in a column and

Figure 35. Action at the ambush site.

the weapons of the ambush force. If possible, armored vehicles are destroyed or disabled by fire of antitank weapons, land mines, molotov cocktails, or by throwing hand grenades into open hatches. An effort is made is immobilize armored vehicles at a point where they are unable to give protection to the rest of the convoy and where they will block the route of other supporting vehicles.

(2) *Ambush of trains.* Moving trains may be subjected to harassing fire, but the most effective ambush is derailment. The locomotive should be derailed on a down grade, at a sharp curve, or on a high bridge. This causes most of the cars to overturn and results in extensive casualties among the passengers. It is desirable to derail trains so that the wreckage remains on the tracks to delay traffic for long periods of time. Fire is directed on the exits of overturned coaches; and designated groups, armed with automatic weapons, rush forward to assault coaches or cars still standing. Other groups take supplies from freight cars and then set fire to the train. Rails are removed from the track at some distance from the ambush site in each direction to delay the arrival of reinforcements by train. In planning the ambush of a train, remember that the enemy may include armored railroad cars in

the train for its protection and that important trains may be preceded by advance guard locomotives or inspection cars to check the track.

(3) *Ambush of waterway traffic.* Waterway traffic, such as barges, or ships, may be ambushed similar to a vehicular column. The ambush party may be able to mine the waterway and thus stop traffic. If mining is not feasible, fire delivered by recoilless weapons can damage or sink the craft. Fire should be directed at engine room spaces, the waterline, and the bridge. Recovery of supplies may be possible if the craft is beached on the banks of the waterway or grounded in shallow water.

Section III. OTHER INTERDICTION TECHNIQUES

91. Mining, Sniping, and Expedient Interdiction Techniques

a. Mining affords the Special Forces and indigenous forces commanders a means of interdicting enemy routes of communication and key areas with little expenditure of manpower (fig. 36). Mines can be used with other operations or used alone. When used alone they are emplaced along routes of communication or known enemy approaches when traffic is light. This allows personnel emplacing the mines to complete the task in secrecy without undue interference. The use of mines to

Figure 36. Enemy sentinels activate mine placed by guerrillas.

Figure 37. Sniping.

cover the withdrawal of a raiding or ambush force slows enemy pursuit and their use in the roadbeds of highways and railroads interferes with movement. Mines can be emplaced around enemy installations and will cause casualties to sentinels and patrols; tend to limit movement; and cause low morale among enemy troops.

b. Sniping is an interdiction technique. It is economical in the use of personnel and has a demoralizing effect. A few trained snipers can cause casualties among enemy personnel, deny or hinder his use of certain routes, and require him to employ a disproportionate number of troops to rid the area of snipers (fig. 37). Snipers may cover an area that has been mined, act as part of a raiding or ambush force, or operate by themselves. Snipers operate best in teams of two, alternating the duties of observer and sniper. Snipers may be used effectively in border denial operations from positions in field fortifications and lookout towers.

c. Expedient interdiction techniques against enemy personnel can be used to the fullest extent for security or denial operations against the enemy. Some of these devices are barbed wire, sharpened stakes, impaling poles, man traps, and all types of boobytraps. See FM 31-73 and FM 5-31.

CHAPTER 8
WATER OPERATIONS

Section I. GENERAL

92. Planning and Preparation

a. Amphibious operations may frequently be employed in guerrilla warfare or counterinsurgency operational areas having exposed coastlines, coastal rivers, or harbors. Infiltration, exfiltration, evacuation, or resupply may be accomplished using amphibious techniques. While water landing sites lack the flexibility of air delivery sites, infiltration operations by both sub-surface and surface craft is considered the most secure and efficient means of providing required support for operational detachments.

b. The landing craft used depends upon the type of ship employed, the tactical and physical conditions at the landing site, the number of personnel, and the amount of cargo to be landed. Of basic concern to the detachment commander is the type of landing craft he will be assigned for operations. Normally, the inflatable reconnaissance boat (designated IB(L) with a capacity for 12 men or IB(S) with a capacity for 7 men) will be used for this purpose. While in the briefing center and before leaving the departure site, the detachment commander insures that his detachment is proficient in amphibious operations.

c. Upon assignment of a transport vessel, the detachment commander should become familiar with the ship's characteristics, facilities, and interior to include exact troop locations and storage areas for the detachment's equipment and boats. Equipment is then prepared and packaged in containers of specific dimensions and weights, according to the requirements of the assigned craft. All equipment is waterproofed and marked for identification. When a submarine is the transporting craft, inflatable boats are deflated and stowed in the free-flooding portions of the superstructure.

93. Tactical Considerations

a. Amphibious operations conducted in support of Special Forces units may be divided into the following phases:

(1) Movement by transport craft to the debarkation point.

(2) Debarkation from transport craft and movement to the landing site in small landing craft.

(3) Disposal of landing craft. In some situations, the detachment may be required to secrete or cache their landing craft for possible use in exfiltration or evacuation from the operational area. Landing craft may sometimes, when specific

mission dictates, be returned to the transport vessel by naval personnel supporting the operation.

(4) Movement of personnel/cargo from the beach site into the objective area and sterilization of the site.

(5) Transferring personnel/cargo from the transporting ship to indigenous craft at a rendezvous point at sea. The indigenous craft, with personnel/cargo concealed or provided with cover, proceed to the landing site in a normal manner.

b. To prepare the detachment for these phases, the detachment commander conducts extensive joint training exercises. Numerous drills in the basics techniques of debarking, boat handling, use of escape trunk, recovery of personnel, and use of communications equipment are accomplished. Complete rehearsals are conducted daily. In addition, if the move is to be made by submarine, physical conditioning exercises should be a part of the daily routine. If a completely submerged transit is made, this will preclude any possible physical exercise because of the restrictive limitations of the submarine.

Section II. LANDING OPERATIONS

94. Debarkation Techniques

Debarkation and ship-to-shore movement are precise operations which can easily jeopardize the safety of all elements and compromise the mission.

Detection of the submarine before debarkation could result in delay, modification, or cancellation of the operation. The choice of a debarkation method will depend largely upon intelligence gathered during the planning phase and on the type of mission, as well as by the degree of training of the Special Forces detachment to be infiltrated.

a. Surface Launch. The surface launch may be conducted by either the wet or dry method. The dry launch consists of the submarine personnel inflating boats, sliding them over the side into the water, positioning them at specified stations, and debarking troop personnel. The wet launch is similar to the dry launch except that after the boats are inflated they are positioned on the deck, Special Forces personnel debark, and the submarine submerges beneath the boats.

b. Broached Launch. This procedure requires the submarine to surface with decks awash. The Special Forces personnel debark through the upper conning tower hatch into the water. As soon as the last swimmer leaves the conning tower, the hatch is secured and the ship submerges at dead-slow speed, preventing propeller wash from injuring the swimmers. The swimmers may then make a compass swim to the beach or inflate the boats in the water, secure equipment, and row the boats to shore. This technique restricts the amount of equipment that can be carried.

c. Bottom Lockout. This procedure can be used when debarking from a submarine and when the vessel can come to a full stop, resting submerged on the bottom. Special Forces personnel, equipped with self-contained underwater breathing apparatus (SCUBA) (open and closed circuits), are locked out in a compartment that can be flooded, allowing them to swim out and make a submerged, compass swim to the beach. This technique can also be used for debarking personnel without SCUBA equipment allowing free, or buoyant, ascent to the surface. Swimmers can then make a compass swim to the beach. It is also possible for a pair of SCUBA divers to "lockout," release, and inflate boats, and for the remainder of the detachment to make a free ascent to the surface with their equipment. After surfacing, they man the boats and row a compass course to the beach.

d. Debarkation from Surface Craft. The detachment debarks by climbing down landing nets into inflatable boats. Because of security, the surface craft remains beyond the horizon and other landing craft delivers the detachment to the beach or tows them to a prescribed release point. Submarines may also surface, release boats on the surface, and tow the inflatable boats by fastening a towline to the aft end of the hangar or conning tower. The boats are towed to a designated release point off the landing site and released.

95. Conduct of Landing Operations

Conducting landing operations on selected landing sites may be accomplished with or without reception committees. When reception committees are present, the supporting naval craft receives the proper identification signals through infrared lights or other visual signaling devices. Landing operations are normally conducted as follows:

a. One hour before debarkation, the Special Forces commander is alerted and all detachment equipment is assembled near debarkation hatches, according to the selected debarkation technique. Upon reaching the debarkation point, the Special Forces commander reports to the conning tower for a periscope view of the beach and to receive last-minute instructions for landing.

b. Once debarked, the control of the team reverts to the Special Forces detachment commander. Navigation is difficult during the ship-to-shore movement. The course to the landing site must compensate for set and drift of known currents. A combination of dead reckoning, celestial observation, and shoreline silhouette is used in navigating. Underwater sound, infrared light, radar, visible light, or radio can be used for navigating if the situation permits. Boat control depends upon the number of boats in the landing party.

c. As the boats approach the surf zone, the detachment commander orders the boats to lie outside the surf zone and to maintain relative position to the beach. Scout swimmers enter the water, approach the beach, and determine any enemy presence in the landing area. This usually is done by moving singly about 50 meters in opposite directions after reaching the beach.

d. After determining that the landing area is clear, they signal the boat team by prearranged signal. This same technique is followed even though a reception committee is present. The scout swimmers will establish security and covering fire until the boat is beached and the remaining personnel secure the equipment. After sterilizing the landing site, the detachment conducts the remainder of its infiltration by land. If a reception committee is available, they will guide the detachment to safe areas where additional equipment that may be brought ashore can be stored, establish security, and assist in sterilizing the landing site.

96. Landing Site Selection

a. Considerations. Normally the landing site selected for detachment infiltration is the responsibility of the supporting service and the SFOB; however, once the detachment has been committed into the operational area and the receipt of additional personnel and supplies are anticipated by water-landing, the selection of the landing site is the responsibility of the detachment commander. His selection must be based on the requirements of both the materials to be received and the availability and size of his reception committee. The detachment commanders must have a basic knowledge of hydrography in order to select sites that will permit transport vessels to conduct landing operations in close proximity. He gives the same consideration to selecting water landing sites as he does to the selection of DZ's or LZ's.

b. Locations. The site should be located to provide maximum freedom from enemy interference. Stretches of heavily defended or frequently patrolled coastlines should be avoided. If sand beaches are used, tracks and other signs of operations that may compromise the mission should be obliterated. Rural, isolated areas are preferred.

c. Exits. The coastal area immediately behind the landing site should be suitable for clandestine landing site and providing a concealed avenue(s) of exit and areas of evasion should be operation be compromised.

d. Barriers. The technique of backward planning, i.e., a step-by-step visualization of the operation commencing with the safe arrival of over-the-beach-infiltration personnel/cargo at the guerrilla base, may disclose that landing sites satisfying landing party and reception committee criteria are unsuitable due to the existence of enemy barriers between the site and the ultimate destination. Such barriers may consist of heavily-patrolled roads; large areas affording little or no concealment; extensive mangrove swamps; areas populated with enemy sympathizers; or an area extensively used for enemy troop maneuvers and training. Unless a secure means of circumventing any existing barrier is devised, the selection and use of landing sites will be limited.

97. Reporting Landing Sites

a. The following data is the minimum required for a beach landing site (additional hydrographic data may be required in some situations):

(1) *Code name.* Procured from the SOI. Designate as primary or alternate (PRI-ALT).

(2) *Location.* Military grid coordinates.

(3) *Description.* The minimum elements of the description are—

(a) Beach gradient at high and low tides.

(b) Metric dimension of usable landing site(s).

(c) Identifying landmarks.

(d) Navigational hazards (shoals, rocks, rips).

(e) Soil condition (sand, mud, shingle, swamp).

b. If, in addition to reporting the landing site, a resupply, evacuation, or exfiltration mission is to be requested, the following additional data are included:

(1) Date/Time mission requested.

(2) Mission requested, e.g., supplies, evacuation, or exfiltration.

(3) Alternate site code name.

c. Data reported on beach landing sites is maintained on file at the SFOB and by the supporting naval unit, similar to DZ and LZ reports. Should a detachment desire to use a previously reported beach landing site, reference is made only to the code name and the date/time the mission is requested.

d. An alternate landing site should be designated and reported in each request for amphibious operations. Prior coordination for use of the alternate site affords flexibility and increases the chances of success. Tide conditions and seaward visibility may not vary much despite widely separated landing sites. Also, an amphibious operation is effected by rate of water movement. For these reasons the times designated for the operation at the primary and alternate landing sites are reported by 24-hours, plus or minus a 30-minute tidal variation.

e. Mission confirmation messages are transmitted to the detachment following the processing of the landing site data and resupply or evacuation request. This message includes the time of arrival required by the reception committee.

f. See appendix III for sample beach landing reports and confirmation report.

98. Marking of Beach Landing Sites

a. Beach landing sites are marked for night operations by means of visible or infrared lights arranged in a predesignated (SOI) code pattern or transmitting a pre-arranged code signal. Panels, arranged in a predesignated code pattern may be used for daylight operations.

b. Landing site markings indicate the proper landing point and serve to identify the reception committee. At night, ship-to-shore identification is accomplished by coded light signals either visible or infrared, and panel signals are used during daylight operations.

c. Identification signals between rendezvousing

ships, such as a navy transport and an indigenous-type craft, are passed with the boat coming from the landing site identifying itself first.

d. The display of beach landing site markers should be visible only to the seaward approach to the site and be elevated as high as possible to achieve maximum range of visibility. Identification signals exchanged between rendezvousing ships should be oral. If lights must be used, the intensity of the signal lamp and the range of transmission is reduced to minimum extent practical. If lights from other sources, such as dwellings, exists along shore, signals lights must differ in color and brilliance.

e. In coastal areas containing shoals, coral reefs, or other navigational hazards, the return trip of the reconnaissance boats or the indigenous-type craft may be facilitated by the temporary anchoring of small, battery-powered lights affixed to bamboo poles at intervals during the outward bound trip. These lights, marking course changes, or hazards, are retrieved as passed during the return voyage. The offshore location of the rendezvous point will be specified for each mission. In principle, it is located to provide the shortest travel of the smaller craft consistent with the security of the transport. Actual locations are dependent upon tides, hydrography, enemy capabilities, normal offshort traffic, visibility, and surface conditions.

99. Reception Committee

The organization of reception committees for

amphibious operations is similar to that for air operations and the same basic functions of command, security, marking, recovery (unloading), and transport must be fulfilled.

Section III. UNDERWATER OPERATIONS

100. General

Because of their training in underwater operational techniques, Special Forces detachments can conduct operations successfully in all operational areas, or near water areas. These techniques involve the use of (SCUBA) equipment, both open circuit and closed circuit systems.

101. Tactical Underwater Operations

a. Rivers, lakes, canals, other inland waterways, and coastal waters adjacent to likely target areas afford excellent opportunities for Special Forces and selected indigenous elements to conduct interdiction missions.

b. Targets adjacent to water generally will be heavily guarded from land attack but lightly defended from water approaches. Special Forces and selected indigenous personnel, adequately equipped, can approach targets underwater and enter the target area, caching their underwater equipment on shore. By appropriate ditching procedures they can also cache equipment underwater if required. Diversionary land attacks by other elements of the attacking force will increase the chances of success by drawing attention away from the water area. After placing their charges, the force can quickly and secretly depart the target area in the same manner they arrived.

c. Other operations using underwater techniques—

   (1) Small-team reconnaissance of harbors, industrial water sites, shipping lanes, mine fields, submarine defenses, docks, and dams.
   (2) Establishment and recovery of underwater caches, pre-positioned or air-dropped.
   (3) Recovery of underwater caches abandoned earlier for emergency reasons.

d. The selection of targets determines the type of equipment and procedures to be employed.

102. Capabilities of Equipment

a. The open-circuit SCUBA system is used in rivers and coastal waterways where obstacles and debris in the water in the target area cause water turbulence and security is not of primary importance. The swimmer can surface safely, cover long distances underwater, and dive to greater depths. With this system, obstacles and mine fields can be checked without fear of compromise.

b. When maximum security is required in entering target areas, the closed circuit SCUBA system is employed. This system does not advertise itself with streams of bubbles like the open-circuit system. Swimmers can thereby approach and work in the target area with less fear of discovery.

103. Limitations

SCUBA equipment, being cumbersome and difficult to maintain, should be airdropped into the operational area as needed. By following this procedure, the equipment will be in a good state of maintenance and safe for use.

a. The open-circuit and closed-circuit systems require special tools and repair kits, as do the hoses, rubber wet suits, and face masks.

b. Closed-circuit SCUBA equipment is particularly difficult and dangerous to operate and only trained personnel should attempt to employ such equipment. Large quantities of baralyme are required to prepare the air tanks with the pure oxygen base for breathing. Since closed-circuit SCUBA uses rebreathable air, the baralyme must be changed after every operation. Dampness and humidity affect this material and render it extremely dangerous.

Section IV. SMALL-BOAT OPERATIONS IN SUPPORT OF COUNTERINSURGENCY

104. General

This section provides guidance for Special Forces units and personnel in the employment of boat-transported combat units in support of counterguerrilla operations on inland waterways, lakes, rivers, canals, and estuaries. Tactics and techniques herein will apply to the employment of motor boats and nonmilitary craft such as sampans, junks, and other indigenous craft. In areas possessing a dense network of inland waterways, small boats can provide a high degree of mobility for Special Forces and indigenous forces in guerrilla operations or operations in insurgent-controlled areas. Small boats are used in much the same role as light trucks to perform a variety of military tasks.

105. Planning Considerations

Special Forces detachment commanders, responsible for planning and conducting operations employing small boats, must consider many factors. Small boats should be considered as a means of transportation and not normally fighting vehicles. Boat patrols, or boat-borne security elements will halt quite frequently and debark small teams to observe or scout the riverbank. In some instances these elements may be required to fight from small boats. This might occur in reaction to an ambush, surprise enemy activity, when pursuing fleeing enemy boats, or when operating in swamps and flooded areas where debarkation is impossible. Detachment commanders must analyze all the advantages and limitations in operating small boats.

a. Advantages.

(1) *Speed.* Motor-powered boats can attain speeds up to 32–37 miles per hour. Even at slower speeds, boat-borne forces can

move more quickly in areas with dense waterway systems than foot troops or armored personnel carriers.

(2) *Obstacles.* Traveled waterways usually are free of obstacles which prevent movement, but shallow-draft boats can maneuver around obstacles as they are encountered. This is especially apparent in dense jungles and coastal mangrove swamps where waterways often constitute the only satisfactory routes of surface movement. Obstacles, such as fallen trees, fish traps, mines, and other man-made or natural obstacles, will slow movement considerably.

(3) *Boat capacity.* Units moving by boat can carry far more weapons and equipment with them than foot elements. Greater combat power can be brought to bear on the enemy; however, in planning boat operations, commanders must insure that adequate means are available to move and use equipment effectively after troops debark or take only such equipment as can be man-made.

b. *Limitations.*

(1) *Canalized movement.* Boats are confined to waterways; they have no cross-country capability. Movement may be restricted by a heavy volume of water traffic; however, boats can be manhandled for short distances overland.

(2) *Concealment and cover.* Some waterways that normally are used as communications routes are devoid of cover and concealment, especially if they are wide. Boats can be seen and fired upon easily in daylight. This may limit boat operations to night movement or limit movement by requiring boats to travel close to the stream banks where shadows and overhead branches aid concealment.

(3) *Noise.* When power driven boats are used the noise of outboard motors reduces the degree of surprise and secrecy required in operations such as ambushes, raids, and infiltrations. The secrecy and surprise required may be obtained by rowing indigenous rivercraft.

(4) *Landing requirements.* Boats must go to or near the shore to unload troops in shallow water or on land. This requirement limits the reaction time of the force in surprise situations such as ambushes. Convenient landing sites are often not available.

106. Special Considerations

Boat operations are not basically different from other operations using special means to increase speed of movement (trucks, helicopters). Backward planning techniques are used. Commanders should avoid limiting the general scheme of maneuver on the basis of available water routes.

Boats are intended to increase, not restrict the range of choices available. Detachment commanders should consider—

a. Intelligence.

(1) Terrain intelligence takes on special importance for the detachment commander when planning small-boat operations. He makes a very careful analysis of the waterway net, especially in enemy-dominated areas. Here the pattern of canals and streams may be altered to suit the enemy and confuse opposing forces. The detachment commander makes extensive use of aerial photographs which present a detailed picture of drainage patterns, new canals, obstacles, and recent changes not readily available on maps. Photos or photomaps are desirable for determining location during movement because of the difficulty in measuring distance traveled on water.

(2) This information is supplemented by records of actual conditions observed while moving on the waterways. Information is recorded concerning obstacles, possible ambush sites, current speed, water plant growth, amount of tidal effect, and other factors that may influence boat operations.

(3) In analyzing the enemy situation, detachment commanders carefully consider the water avenues available to the enemy

force. Special attention is given to locating and neutralizing boat resources, especially powered craft, that can be used by enemy units. From his analysis and information concerning the waterway, a detachment commander can effect operations to position friendly units to block all water routes available to the enemy.

b. Planning. Detachment commanders participate in the early planning for any operation. On receipt of the initial order, Special Forces detachment commanders will coordinate with support elements and other combat forces their plans of execution, tentative formations and march orders, number of boats available, departure times, embarkation sites, communications, fuel supplies, and rehearsal plans.

c. Communications. A boat-transported force operates its radios within normal tactical nets, whether embarked or debarked. The radios of the boat unit are used for internal control while moving or for communications with higher headquarters when operating independently. Organic radio and other communications equipment, presently in the hands of the Special Forces detachment not compatible to boat operations will be supplemented by additional radio equipment.

d. Supply and Maintenance. Due to the high fuel consumption rate of outboard motors, gasoline supply must be carefully planned. The standard outboard motor requires constant and proper maintenance to insure dependable performance. Detachment commanders should plan for extra maintenance personnel to be assigned to his operational areas. Spare parts and extra motors may be taken when operations are conducted away from the main base for extensive periods of time.

e. Organization. The detachment commander remains flexible in organizing his small-boat elements. Each mission assigned will dictate the degree of organization required, and the additional equipment and personnel to support his mission. Additionally, the scheme of maneuver in the objective area may well dictate the size and composition of the force as well as the order of march and the number of boats required.

107. Tactics

a. General. When Special Forces detachments have organized paramilitary and other indigenous forces into effective fighting units, requirements may be placed on them to conduct operations on inland waterways, lakes, rivers, and canals in or near their operational areas. Special Forces-directed, indigenous forces may successfully conduct small-boat operations, such as—

    (1) Reconnaissance patrols.
    (2) Ambushes.
    (3) Raids.
    (4) River crossings.
    (5) Movement of supplies.

(6) Support of offensive operations.

(7) Infiltration and penetration operations.

b. Reconnaissance Patrols. Reconnaissance patrols normally use at least two boats which furnish mutual support. They may move by successive bounds, alternate bounds, or continuous movement. The tactics employed are identical to those of motorized reconnaissance patrols.

(1) *Moving by successive bounds.* The boats in the patrol keep their relative place in the column. Boat number 1 moves ahead while the second boat debarks its troops to observe. When boat number 1 reaches a secure position, troops debark and observe until boat number 2 moves up to take over this position (fig. 38).

(2) *Moving by alternate bounds.* The two boats alternate as lead craft on each bound (fig. 39). This method, although more rapid than successive bounds, does not allow personnel in the second craft to observe carefully before they pass the halted lead craft.

(3) *Continuous movement.* All boats move at a moderate speed, maintaining position and security by careful observation. The leading boat stops to investigate areas that seem particularly dangerous with the remainder of the craft maintaining their positions. This is the fastest, but least secure, method of movement.

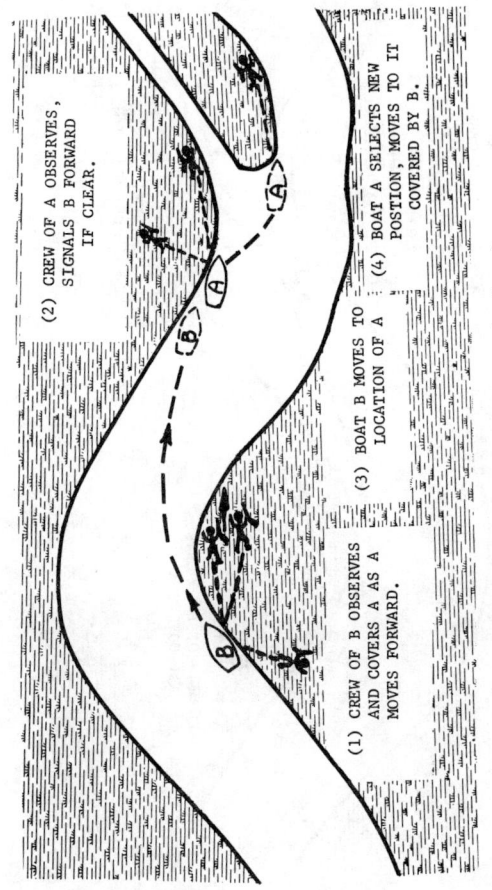

Figure 38. Movement by successive bounds.

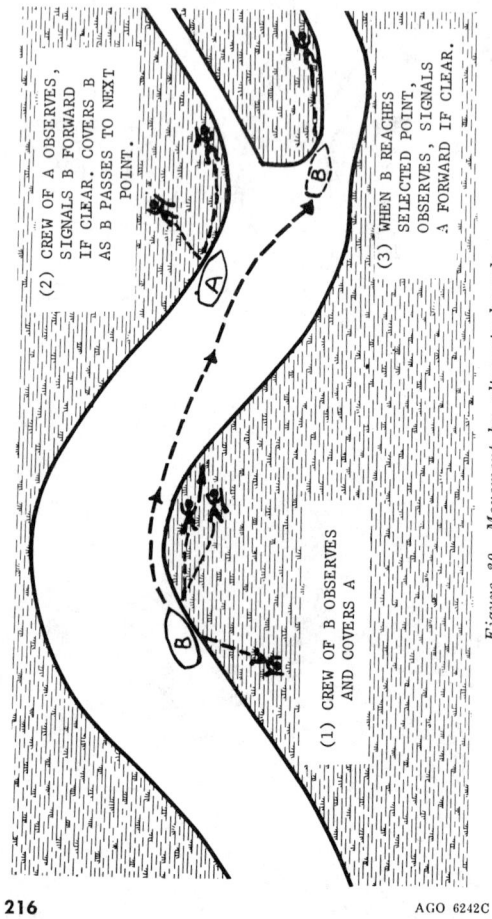

Figure 39. Movement by alternate bounds.

(4) *Radio.* Each boat is provided with a radio for constant contact.

(5) *March security.* The techniques described above, in addition to applying to reconnaissance missions, also apply to Special Forces directed elements assigned as march security (advance guard, flank guard) in support of a larger force on offensive operations.

c. Ambushes. In planning and considering ambushes, the detachment commander first considers the waiting time usually required at the ambush site. During the waiting period, changes in level and direction of stream flow often occur. The commander must anticipate these changes and plan his ambush around them. Changes in the water level due to tides may require relaying weapons in a waterway ambush. At ebb tide boats may be stranded or some withdrawal routes may become too shallow to use. The direction of approach of enemy boats may be based on the direction of current flow. These factors are paramount in choosing the location, timing, and method of ambush.

(1) Certain techniques may be used in small-boat operations that can enhance the Special Forces detachment commander's chances of success during the ambush. Stealth in movement to position is paramount and can be achieved by using oars or poles to propel the boats, instead of motors, when approaching the ambush

site. Boats can drift to position with the current or tidal flow; however, when rowing or drifting to position, motors remain affixed to the boats and in the up position for immediate use in the event of premature enemy contact.

(2) Assuming that the force is under constant enemy surveillance, the detachment commander will consider leaving small ambush parties behind when patrols stop and dismount to observe or reconnoiter specific targets. The ambush is set in the vicinity of the debarkation point, where the boats are left in concealment. This technique is useful if the boat force operates with frequent halts and debarkations.

(3) Whether the ambush is laid to cover a road, trail, or waterway, the force will normally debark and take up concealed positions. Boat crewmen remain in or near their craft which are carefully concealed. During the occupation of the ambush site, the ambush commander assigns responsibility for security of the boats to the crewmen. The crewmen further act as a covering force when the ambush force withdraws.

d. Raids.

(1) The detachment commander may conduct a raid on a specific target wherein he will employ water-borne forces; how-

ever, he must analyze enemy security and defenses very carefully. In the raid, the detachment commander may use his boats to:

(a) Move troops to the line of departure.
(b) Position blocking forces.
(c) Aid in placing or displacing crew-served weapons.
(d) Patrol flanks of attacking unit.

(2) If the objective is some distance from the shore, a boat-borne party may debark and approach stealthily on foot. This technique permits a more thorough search of the area; and, in the event of contact, friendly fire and maneuver will be faster and more effective.

(3) When the objective is near the shore, the assault force may use motor-driven boats to storm directly into the objective area. Amphibious assault tactics are used. The storming tactics should be used only when surprise can be achieved and when the waterway is large enough to permit evasive action by the assaulting craft.

(4) When planning such attacks, smoke from mortar shells and artillery fire, if available, are planned to cover the withdrawal of the raiding force. In this instance, the use of ambush tactics may be employed against a pursuing enemy force since, in all probability, he will be re-

stricted to waterways for speed of movement.

(5) When targets are relatively close to river banks or canals and the enemy has concentrated his security in this area against boat attack, a diversionary attack from the land, conducted in force, may permit Special Forces personnel and trained indigenes to approach and enter the target area using SCUBA gear to place underwater charges and destroy selected facilities.

e. Other Operational Techniques. Small boats may provide the mobility which permits superior fire power to be brought to bear against an insurgent force. The load-carrying capability, maneuverability, and speed of powered craft gives the Special Forces units additional advantages for other types of operations. The detachment commander, may employ his craft in support of civic action programs and environmental improvement programs.

(1) *Medical evacuation.* Small boats provide a relatively fast means of evacuating casualties and sick and injured. When boat evacuation is planned, aid stations may be located for convenient access to the waterway net. Land sites are prepared to facilitate prompt and gentle unloading of evacuees.

(2) *Movement of supplies.* When operating in remote areas, especially during ad-

verse weather conditions that affect air operations, small boats can assist in all forms of logistical supply and movement. This transportation may include ammunition resupply, fuel transportation, medical supplies, building supplies, food, clothing, and equipment needed for villagers. When necessary, selected engineer equipment such as power tools to aid in building programs may be transported.

(3) *Resources control measures.* When on patrols, the speed of small boats allows them to overtake all indigenous river craft normally used in commerce or employed by the insurgent. Special Forces small-boat operations may be used in policing waterways and searching suspected craft as part of a resources control campaign. Boats can tie-up in critical areas and establish check points in conjunction with local law enforcement agencies.

f. Use of Local Craft. Sampans and other local boats may be used to perform the missions and tactics as described above. Special Forces personnel will be selected and trained to operate the boats in conjunction with the local indigenous force. Local boats may be used for infiltration and disguised movements. When so used, boats are operated in accordance with local practice and custom, and obvious military formations and boat tactics are avoided.

CHAPTER 9
COMMUNICATIONS

Section I. UNCONVENTIONAL WARFARE

108. Systems and Techniques

The communications systems and techniques employed by Special Forces are applicable to both unconventional warfare and counterinsurgency operations. Communications within friendly territory, between friendly territory, and within the GWOA will be discussed.

109. Extent and Type of Communications

a. Communications Within Friendly Territory. Communications between the SFOB and other headquarters or activities in friendly territory generally are the same as those required by any headquarters of comparable size. The facilities of the theater army area signal system are used. When backup or special circuits are necessary they are provided by radio or radio-teletype operated by the Special Forces group. Communications in this area present no unique operational or technical signal problems.

b. Communications to and from the GWOA. When a detachment is committed, the primary means, and often the only means, of communications with the SFOB is radio; but other methods such as infiltration of couriers, exchange of messages during resupply, or other existing communication facilities may be used where practical.

c. Communications Within the GWOA. As a general rule, communications within the GWOA progress from clandestine to overt systems as the guerrilla movement gains strength. The extent and type of system depend on factors such as size of the area, the size of the guerrilla force, activities of the enemy and the guerrillas, the technical proficiency of both the enemy and the guerrilla communication organization, and the required speed of response to the orders of the area command. Any and all means which satisfy the requirement for communications and provide the required security are used. Certain clandestine communications systems may be used, but these should be tightly controlled by the commander (see FM 31–20). All the following are considered—
   (1) Messenger.
   (2) Radio.
   (3) Telephone.
   (4) Audible signals.
   (5) Visual signals.
   (6) Local communication systems.
   (7) Pigeons or trained animals.

110. Communication Media

a. Messenger. In the early development stages of a GWOA, messengers may be the only secure means of communication. In the GWOA, messenger (courier) service is organized using clandestine communications techniques described in FM 31–20. As the GWOA is organized and equipped, security remains a paramount consideration; therefore, communication means will be dictated by the status of training and the capability of the guerrilla force.

b. Radio. Radio can provide instantaneous, generally reliable communications; however, any radio transmisison is vulnerable to interception and jamming by an enemy. The advantage of its speed must be balanced against the probable loss of security. Low-powered, frequency-modulated radios operating in the VHF or UHF band can be used, under some conditions, with little risk. Generally, when considering the use of radio, the deciding factors are the nature of the message text and the probable enemy reaction time if the message is intercepted. For example, enemy reaction to last-minute control instructions during a raid or ambush would not be rapid enough to affect the operation. On the other hand, the interception of plans or instructions involving future actions could result in disastrous compromise. Within a GWOA the availability of radio equipment may be the governing factor. Maintenance, spare parts, and resupply of batteries are important considerations. The use of even the simplest radio requires training, and operators and maintenance personnel must be included in the training program.

c. Telephone. In the early stages of development of a GWOA, telephones may be used extensively, such as between a security outpost and a base camp, or during an ambush to warn of the approach of a convoy or train. When using a telephone under these conditions, it is often advantageous to use a ground-return circuit, allowing the telephones to be operated with a single metallic conductor connecting them. A section of barbed wire fence, unused power line, unused telephone line, or one side of a railroad track can be used as the conductor and may already be in place. The conductor must be insulated from the ground and the other terminal of the telephone must be connected to a good ground connection (fig. 40).

d. Audible Signals. Audible signals are useful only for short distances. Church bells, vehicle horns, musical instruments, sirens, dogs barking, or voices may be used as audible signals. Quite often, audible signals can be planned in such a way that the sound is routine and recognizable as a signal only to someone trained in the system.

e. Visual Signals. Visual signals are limited only by the imagination of the person planning the signals and by the equipment available. Visual signals include—

  (1) Signals sent by light from a flashlight or powerful searchlight or by using sunlight reflected from a mirror. The use of any flashing light requires some prearranged code.

  (2) Signals as simple as having a housewife

Figure 40. Expedient ground-return circuits.

hang laundry on a clothesline to serve as a warning. Light, smoke, a fire, or a person walking over a given road at a specified time can be a visual signal. Normal actions are the guide for developing visual signals.

(3) Flags used to transmit messages either by means of semaphore or wigwag. In semaphore two flags are used. The position of the flags designate a certain letter. Wigwags can be used to send a message by Morse code. The flag on one side of the body indicates a dash, on the other side a dot (see FM 21–60).

f. Local Communication Systems. Many areas of the world have extensive, local communication systems. Without any special equipment, part or all of these systems may be used. When considering the use of these local communications systems, security must be paramount. The local language or dialect must be used in apparently innocent conversation.

g. Pigeons or Trained Animals.

(1) Homing pigeons, obtained locally or from the SFOB, may be used for the rapid, secure transmission of messages within the operational area. Since they require a few days to acquaint themselves with the home loft area, homing pigeons should be used when the guerrilla base is relatively static. Extremely cold weather limits the use of pigeons.

(2) Locally-procured, trained animals (usually dogs) may also be used as a means of communication; however, dogs are usually more susceptible to interception or diversion than homing pigeons.

111. Communication Training

a. General. Radio personnel assigned to Special Forces operational detachments are confronted with problems different from those faced by radio operators assigned to a conventional military unit. When committed to a GWOA, operators must be able to communicate over long distances, up to 4,000 km, using low-powered equipment. They must do this in a manner that will result in a minimum loss of security. Technical assistance and maintenance support are not readily available. Messages are encrypted using pencil and paper rather than machine systems. The radio operators must also be prepared to assist and advise the detachment commander on any communication problem within the area to include the communication training of the guerrilla force.

b. Code Speed and Procedures. A Special Forces radio operator must be able to transmit and receive Morse code at the rate of 18 words per minute. He must be thoroughly familiar with radio-telegraph procedure as described in ACP-124B. Once these standards have been achieved, they must be maintained by constant practice. Before infiltration, the SOP is established for the actual radio-telegraph procedure to be used in the operational area. Sufficient time must be allocated for radio operators to become familiar with this specific procedure.

c. Maintenance and Use of Equipment. Within a GWOA normal maintenance support is not available. In the A or B detachment, any repair of signal equipment is done by the operator or, when feasible, by friendly members of the local populace. Radio operator training includes sufficient theory and practice so that the operator can perform direct support maintenance on the primary detachment radio set. He is sufficiently schooled in theory so that he can make sound recommendation on the use of enemy equipment captured within the operational area.

d. Radio Propagation. The radio frequencies to be used between the GWOA and the SFOB are contained in the detachment's Signal Operational Instructions (SOI). These radio frequencies are determined before infiltration on the basis of published radio frequency prediction charts and tables. Detailed information on selecting frequencies for long-range comunications can be found in TM 11-666 and radio propagation charts which are procured from the U.S. Army Signal Radio Propagation Agency, Fort Monmouth, N.J. 07703. These charts are published monthly and must be requested for the particular area of operations. Conditions of radio wave propagation may vary from those predicted. When this happens, communications between the detachment and SFOB may be impossible for periods of time extending from several hours to a week or longer. The enemy may be able to intercept signals in areas where they normally would not be heard. There is no way in which the radio operator can determine that radio wave propagation is differing from that predicted; however, if he understands radio frequency prediction, he can more readily recognize this situation when it does exist. Frequencies are changed, on order of SFOB, to overcome difficulties caused by changing propagation conditions.

112. Antennas

Special Forces radio operators use field expedients to insure reliable communications. Because of rigid limitations on size and weight of equipment, the radio used by Special Forces is not issued with a prefabricated antenna. Only antenna wire is issued. Although there is little the radio operator can do to increase the designed power output of his transmitter, he can maximize the propagation of his signal by efficient use of an antenna system. Antenna theory and construction are presented in FM 24–18 and TM 11–666. The Special Forces radio operator must understand the material covered in the manuals in order to provide long-range communication. Various types of antennas which can be used with Special Forces-issued radio equipment are shown in figures 42–53.

a. Field Expedient Insulators (fig. 41). When constructing an antenna, it is important to in-

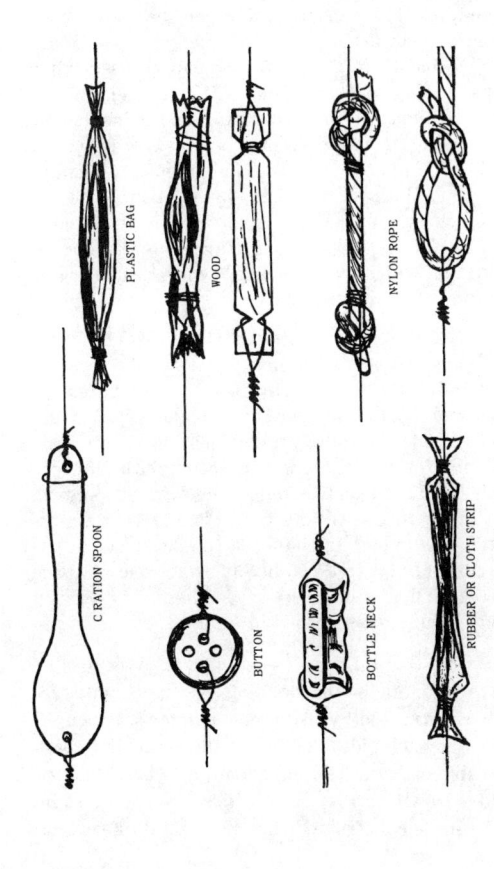

Figure 41. Expedient insulators.

sulate the antenna from its supports or from the ground. It is often necessary for the radio operator to make use of whatever materials are available. Almost any kind of wire can be used when constructing an antenna. Although glass and porcelain may be the best materials for insulators, it is better to use a second best (such as wood) rather than none at all. The antenna diagrams shown in this manual (figs. 42–53) cannot be understood without a basic knowledge of antenna theory. These diagrams picture antenna configurations which can be used with issued radio equipment in limited space.

b. Quarter-Wave Antenna (fig. 42). The quarter-wave length antenna is normally erected vertically. Its length (in feet) is computed by dividing 234 by the operating frequency in megacycles. It is omni-directional, making it an ideal antenna for an NCS when operating with different teams and the exact team locations are not known. It can be used with any type of radio and is normally used when a ground wave is desired. In the case of standard FM radios it makes use of space waves. When a quarter-wave antenna is used, a good ground system should be used.

c. Half-Wave Doublet Antenna. A typical half-wave antenna is the doublet, or dipole antenna. It is constructed by using one-quarter-wavelength wire for each side and fed in the center by coaxial cable or, as a field expedient, a twisted pair of field wire. It can be used with any type of radio and can be constructed in a horizontal or vertical

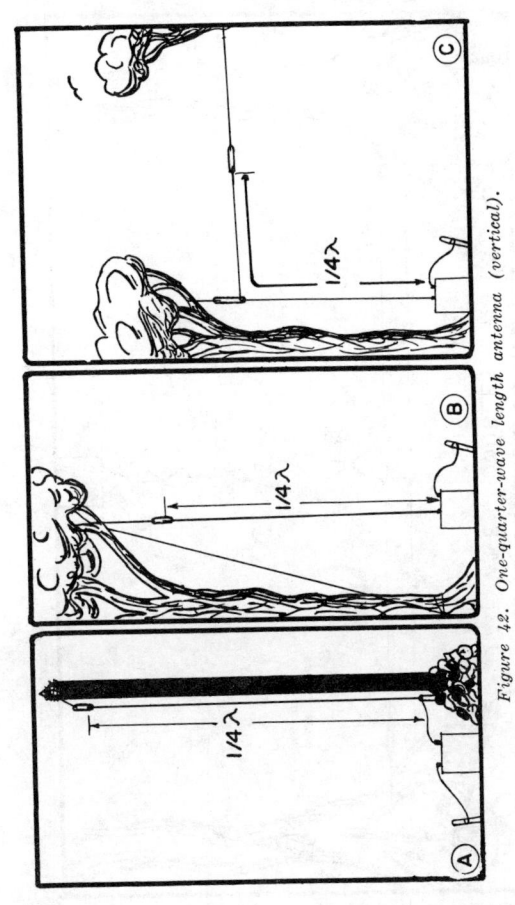

Figure 42. One-quarter-wave length antenna (vertical).

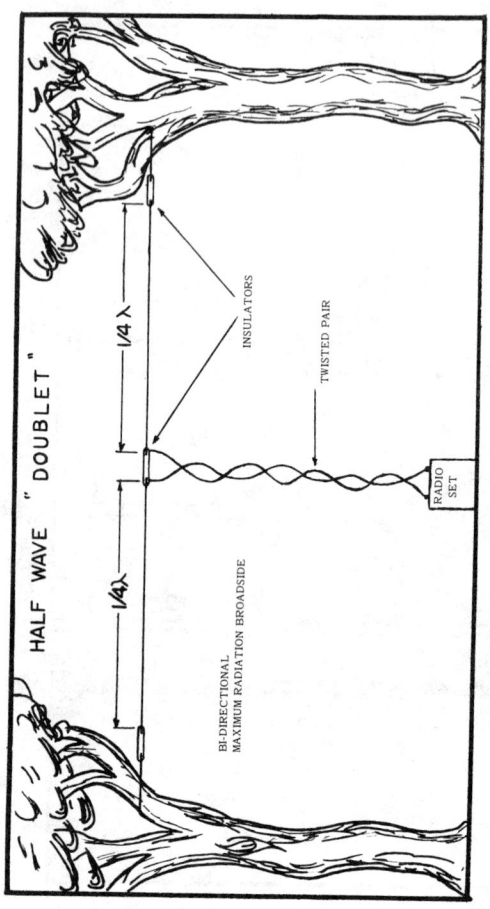

Figure 43. Half-wave doublet antenna.

plane. When in a horizontal position (fig. 43) it radiates broadside or at a 90 degree angle from the antenna. When it is constructed in a vertical plane (fig. 44) it has a radiation pattern of 360 degrees. This antenna is superior to the quarter-wavelength antenna. When connecting this antenna to the radio set, one lead goes to the antenna binding post; the other goes to the ground binding post. No additional ground is necessary.

d. Slant-Wire Antenna (fig. 45). The slant-wire antenna is an efficient radiating system using only a single antenna support. Two pieces of wire, each one a quarter-wavelength long, are used to make up the antenna. One piece is slanted down from the antenna support at an angle of 30 to 60 degrees; it is connected to the antenna post on the transmitter. The other wire is used as a counterpoise just above the ground and laid out from the transmitter away from the slanting wire. If the wire used as a counterpoise is not insulated, it must be insulated from the ground; the counterpoise is connected to the ground post. Maximum radiation occurs in the direction of the counterpoise including the slanting wire.

e. Fourteen Percent Off-Center Fed Antenna (fig. 46). In the event no suitable transmission line is available such as coaxial cable, or twisted pair, a suitable antenna can be constructed using an antenna one half-wavelength long and feeding it with a single wire at a point 14 percent of a wavelength from the center. This antenna is suitable for use with radios such as the AN/GRC–

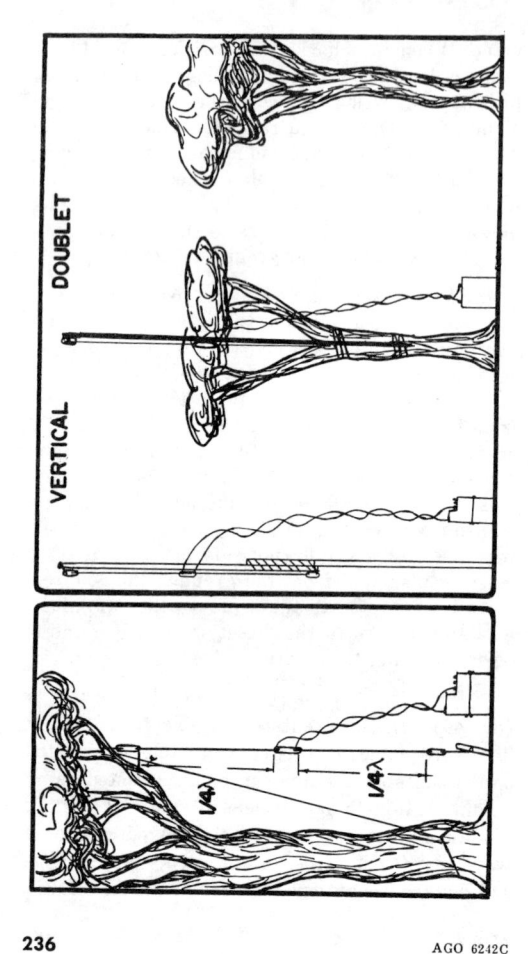

Figure 44. Vertical doublet.

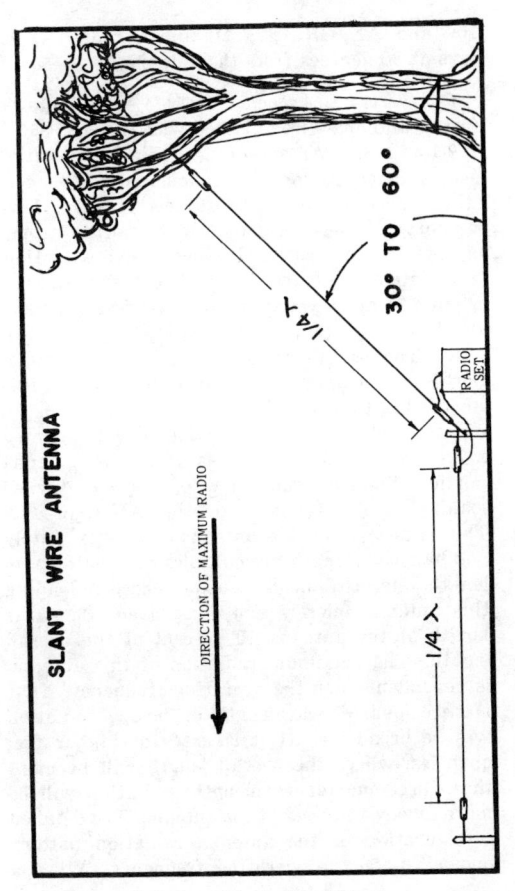

Figure 45. Slant-wire antenna.

109 and AN/GRC–87. Maximum radiation occurs at 90 degrees from the antenna.

f. Inverted L Antenna (fig. 47). One hundred feet of antenna wire is issued with the AN/GRC–109 Radio Set. An antenna suitable for operation over a wide range of frequencies can be constructed using the entire 10 feet of wire. Two supports are selected which will give the proper orientation in relation to the receiving station. The antenna is erected so that the horizontal portion is approximately 60 to 70 feet long and the lead-in portion approximately 30 to 40 feet long. The lead-in must come off the horizontal portion of the antenna at a 90 degree angle. If it is impossible to raise the antenna 30 feet in the air, the antenna lead-in portion may be slanted away keeping it 90 degrees from the antenna portion. Some difficulty may be experienced when loading this antenna with the AN/GRC–109. This is true when the antenna is approximately one half-wavelength or multiples of a half-wavelength long. If difficulty is experienced in loading this antenna, merely add or subtract from the length of the antenna 10 percent of the overall length. The maximum radiation of this antenna is dependent upon the operating frequency. If it is three-quarter-wavelength or below, radiation will be broadside. If it is used for higher frequencies, where the overall length will be more than three-quarter-wavelength, radiation will be more towards the end of the antenna. The detailed configuration of the antenna radiation pattern depends upon the operating frequency. When a

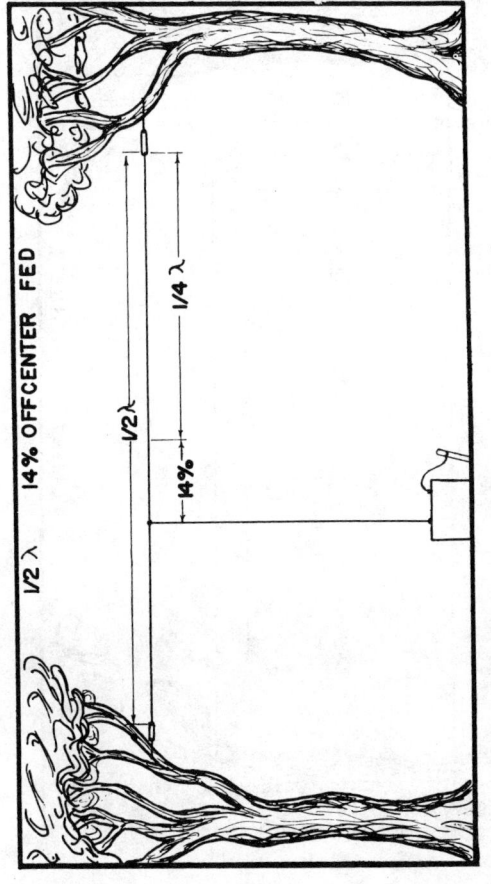

Figure 46. Half-wave antenna-off-center fed.

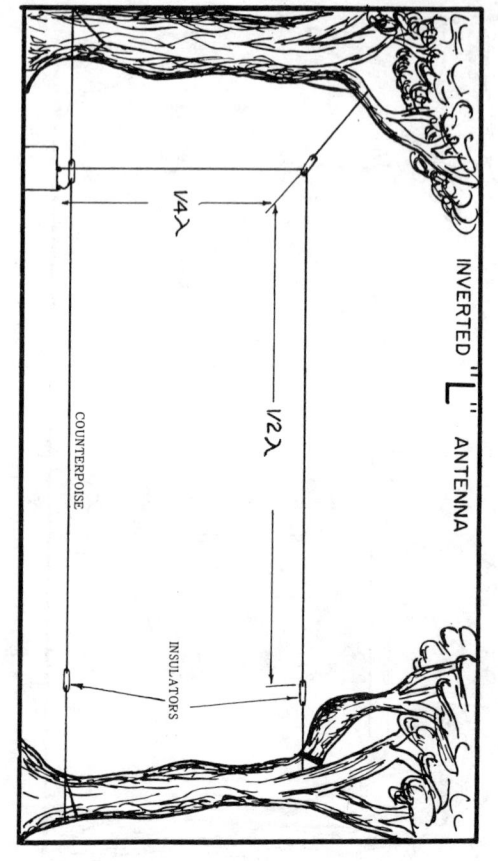

Figure 47. Inverted "L" antenna.

suitable ground is not available for an antenna, a counterpoise should be used.

g. Indoor Antennas. There are times when a Special Forces radio operator must operate from inside a building. When this is necessary, a suitable antenna can still be constructed. Any of the antennas mentioned in this chapter can be used if there is space available inside the building. If space is limited, a loop antenna may be constructed (fig. 48). This antenna is a full

Antenna Length: Full wave
Frequency Shown: 25 megacycles
Normal Range for
Frequency Shown: Day, 750 miles or greater; Night, Frequency too high.

Antenna Length = $\frac{984}{25}$ = 40 ft

$\frac{300}{25}$ = 12 meters

Note: Tune output carefully by indicator lamp. Bulb will not glow brightly.

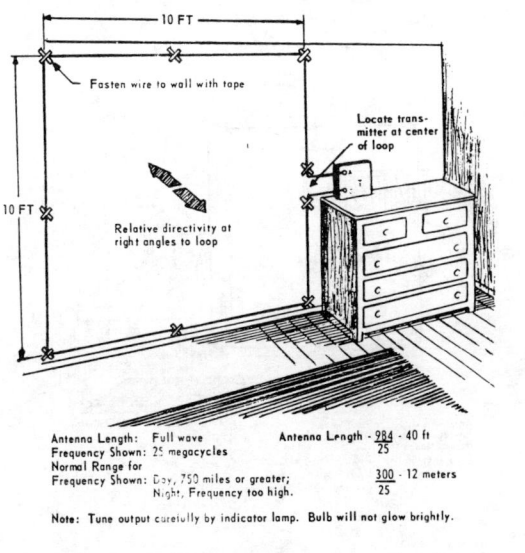

Figure 48. Full-wave square-loop antenna.

wavelength long and is fed directly in the center. It is limited to frequencies whose wavelengths will not exceed the dimensions of the room. For operation on lower frequencies a half-wave, square-loop antenna (fig. 49) may be used inside a building. Excellent results may be obtained if care is taken in constructing and tuning the antenna. This is important when operating the AN/GRC-109 since the indicator lamp of the antenna will not glow brightly with either the full-wave loop or the half-wave open loop. Al-

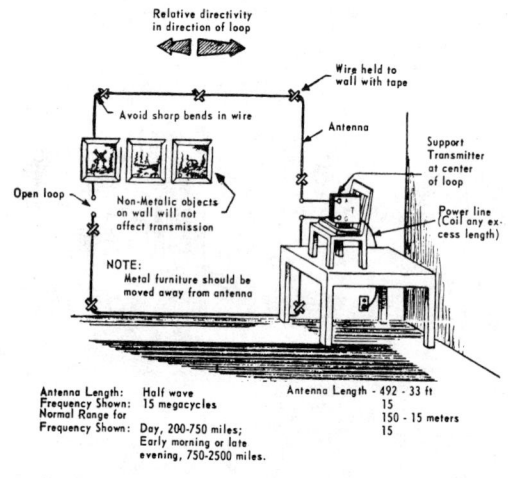

Antenna Length: Half wave
Frequency Shown: 15 megacycles
Normal Range for
Frequency Shown: Day, 200-750 miles;
Early morning or late evening, 750-2500 miles.

Antenna Length - 492 - 33 ft
15
150 - 15 meters
15

Note: Tune output carefully by indicator lamp. Bulb will not glow brightly.

Figure 49. Half-wave square-loop antenna.

though these antennas may be used indoors, it must be remembered that best results are obtained when operating with an outdoor system.

h. Jungle Antenna (fig. 50). In both the unconventional warfare and counterinsurgency roles, it may be necessary to have patrols operating outside the normal range of FM radio sets. When this is necessary, an antenna system can be constructed which will allow communications beyond the normal range of current radios. This can be accomplished through the use of the jungle (fig. 50), the wave (fig. 51), the balloon (fig. 52), or the half-rhombic antennas (fig. 53). When operating on frequencies above 30 mc, the transmission range can be increased by improved antenna. The use of any one of these antennas more than doubles the range of standard FM radio sets.

113. Message Writing

a. The writer of a message must express his thoughts clearly and concisely. Additional transmission time caused by unnecessary message length gives the enemy a better opportunity for interception and radio direction-finding, and furnishes more traffic for analysis.

b. The following basic rules are applied to all messages:

  (1) *Preparation.* All outgoing messages to the SFOB are prepared or reviewed by the detachment commander or his executive officer before transmission.

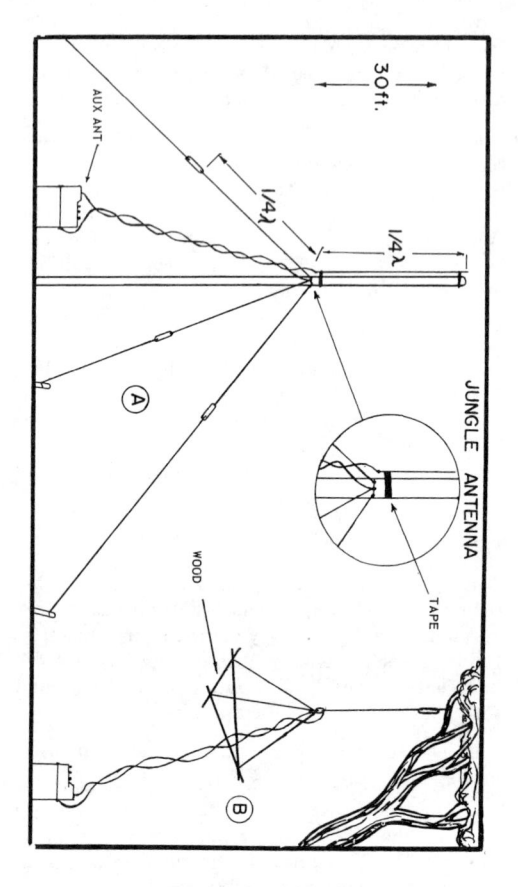

Figure 50. Jungle antenna.

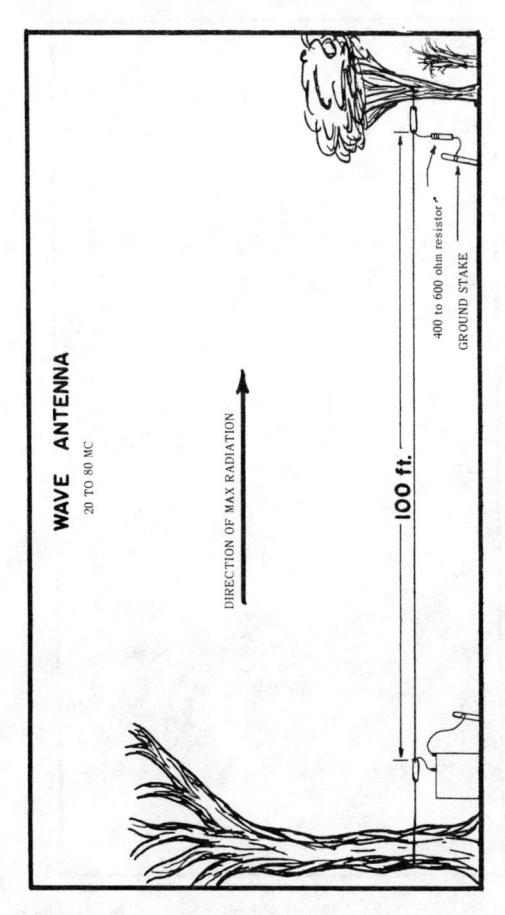

Figure 51. Wave antenna.

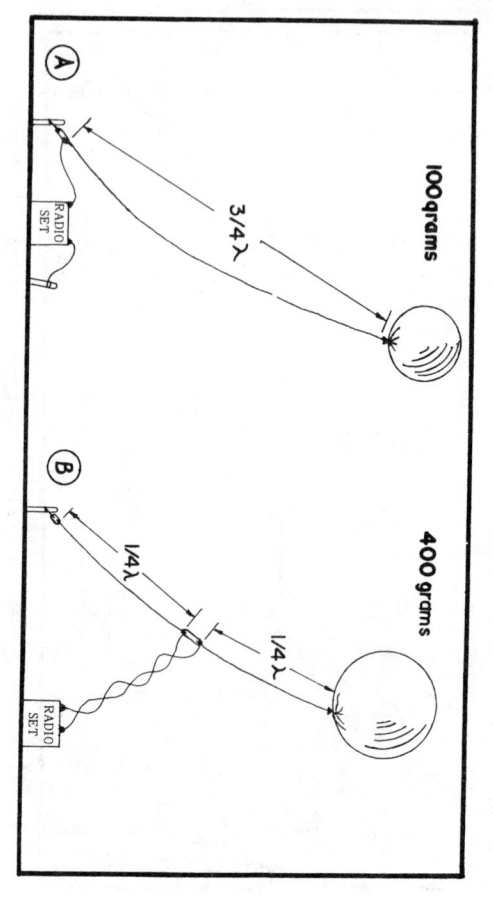

Figure 52. Balloon-supported antenna.

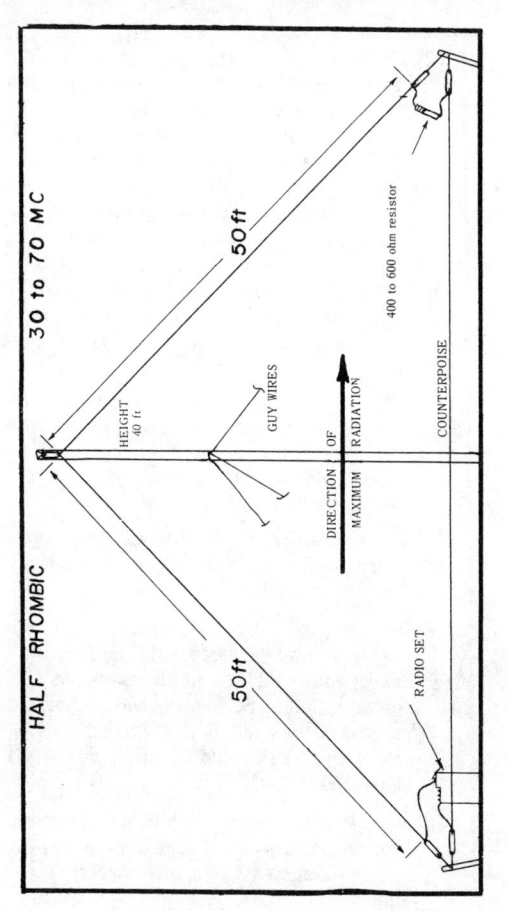

Figure 53. Half-rhombic antenna.

(2) *Content.* Write the message and then read it back. First consider any portion that can be eliminated. Many times the bulk of a message is used to say something that is obvious by the very fact that the message is being sent. Consider each portion. Does each portion tell the addressee something, or could that whole sentence or thought be eliminated? Once this has been done, consider whether the thought of the message is expressed clearly and concisely as possible.

(3) *Writing.* Print carefully to avoid any confusion about the meaning of the message. An encrypted message may be made completely useless by one misunderstood letter.

(4) *Abbreviations.* Use authorized abbreviations and only when they will not be misunderstood.

(5) *Punctuation.* Do not punctuate unless necessary for clarity. Do not use the expression STOP in a message. If punctuation is necessary, use authorized abbreviations such as QUES, CLN, PAREN, PD, CMM, PARA, AND QUOTE-UN-QUOTE.

(6) *Repetition.* Repeat only to avoid errors, not for emphasis. For example, repeat unusual names to insure correct spelling.

(7) *Numbers.* Numbers may be written as digits or spelled out. When spelled out, they are expressed in words for each digit except in exact hundreds or thousands, when the word hundred or thousand is used. Some cryptographic systems require the numbers to be encoded without spelling. As a general rule, numbers should be spelled out before encrypting. If the message is completely understood the first time it is transmitted, the result will be less time on the air. Example: 123.4 is written as ONE TWO THREE POINT FOUR; 500 is written FIVE HUNDRED and 20,000 as TWO ZERO THOUSAND.

(8) *Isolated letters.* If necessary to use isolated letters, use the phonetic alphabet for each isolated letter.

 c. Codes are normally used for brevity. Extensive brevity codes can be developed by proper planning which can greatly enhance message brevity and clarity. Codes that may be employed by Special Forces detachments in their operations are—

(1) The Catalog Supply System (CSS) provides an operational detachment with a brevity code in which single or several associated logistical items may be requested on resupply operations (see sample Catalog Supply System, app. VII).

(2) The Q and Z signals used by radio operators (ACP 131).

(3) Operations codes (SOI).

114. Communication Security

a. Security is of particular importance to a Special Forces detachment located in a GWOA. A violation of any of the principles of communication security endangers the detachment. Communication security is the protection resulting from all measures designed to deny unauthorized persons information of value which might be derived from a study of communications. Communication security is obtained through proper physical security, transmission security, and cryptographic security.

b. Physical security is defined as that element of security which results from the physical measures taken to safeguard communications documents, equipment, and personnel. Within the SFOB, physical security measures are similar to those of any military organization.

(1) Operational detachment personnel obtain security clearance before infiltration. The detachment commander must use his judgment and discretion in dealing with indigenous personnel and allowing them access to classified information. Information on cryptographic systems used by Special Forces is never released to indigenous personnel.

(2) Classified material is kept on the person of one of the detachment members or under constant guard. The physical security of the radio set is maintained by choosing good transmission and storage locations and by having a minimum number of persons know these locations. Techniques of physical security applicable to Special Forces in GWOA are—

  (a) Avoid tops of mountains. The enemy will search the most obvious spots first. If technically feasible, locate the detachment radio transmission site on a forward slope or in a valley.

  (b) Move the radio after each transmission.

  (c) Sterilize radio sites.

  (d) Place surveillance on radio sites before and after transmissions.

  (e) Post guards when waiting for, and during actual transmission.

  (f) Do not construct an antenna over a path. Do not use shiny antenna wire.

  (g) Do not carry classified material to transmission site.

(3) Detachment cryptographic systems and SOI's must not fall into enemy hands. Care must be taken not to destroy these items prematurely since replacement is difficult. Remember, however, that destruction by burning is not complete unless the ashes are destroyed.

c. Transmission security includes all measures designed to protect transmissions from interception, traffic analysis, direction finding, and imitative deception. Some techniques of transmission security applicable to Special Forces operations in a GWOA are—

(1) Make minimum transmissions.
(2) Do not tune transmitters until exact contact times.
(3) Locate transmitter sites so that known direction finding stations are beyond ground wave distances.
(4) Transmit on an irregular schedule.
(5) Never transmit from the same area twice.
(6) Send short messages.
(7) Use highly directional antennas.

Many times it may be necessary in the interest of transmission security to compromise between technically favorable transmission sites and transmission sites which meet the physical and transmission security criteria outlined above.

d. Cryptographic security results from the proper use of technically sound cryptographic systems. Systems and means available to the SFOB and the detachment commanders will vary with missions and operational areas. Specific instructions, techniques, and methods to be used are covered in premission briefings on a need-to-know basis.

Section II. COUNTERINSURGENCY

115. Systems and Techniques

a. Initial Requirements. In establishing an SFOB in a counterinsurgency environment the same signal considerations apply as in an unconventional warfare environment. The area for counterinsurgency operations is considered to be friendly territory in the same meaning used in unconventional warfare. Under the operational control of the Special Forces group signal officer, the signal company will establish the following communications as directed:

  (1) Installation, maintenance, and operation of an internal wire system for the SFOB.
  (2) Termination of landline circuits from higher, adjacent, and lower headquarters.
  (3) Entry into radio nets of higher and adjacent headquarters as required.
  (4) Installation, maintenance, and operation of an appropriate cryptographic facility and communications center.
  (5) Installation, operation, and maintenance of communications to subordinate units.

b. Control Requirements. As the Special Forces effort develops in the area, provisions must be made for communications to be deployed detachments subordinate to the SFOB. Normally, the chain of command will be used for the chain of communications: C detachments will be responsible for communications to their deployed B detachments; B detachments to their deployed A detachments. Detachment commanders will establish their own internal communication systems.

c. Signal Company. The signal company will be employed in a slightly different role from that in unconventional warfare. Radio teletype teams, or sections of the mobile radio platoon, will be permanently deployed within the C detachment headquarters. These teams, while operating under the operational control of the C detachment commander, will necessarily remain under the command of the signal company commander. Only necessary personnel and equipment of the base operations platoon and the two-base radio platoons to operate SFOB communications will remain at the SFOB. The balance of the personnel and equipment will be used to augment the C detachments and, where necessary, the B detachments.

d. Radio Communications. In a counterinsurgency environment, longline communications are not normally installed; primary reliance is placed on radio communications. The TOE radio equipment of the Special Forces detachment is not suited for the high-volume, encrypted traffic load required. Some of the types of radio nets required are:

(1) *Command message nets.* Radio teletype, encrypted (on-line or off-line) to handle operational, logistical, and administrative message traffic.

(2) *Command voice nets.* Voice operated,

radio-telephone radio nets using suitable, available, AM-voice or single side band radio equipment. The primary purpose is to provide the commander and staff direct contact with appropriate personnel in subordinate headquarters. Radio wire integration is incorporated at all command levels of these nets whenever possible.

(3) *Emergency Nets.* An emergency net will use both TOE Special Forces radio and other radio equipment provided to augment the group for its counterinsurgency mission. The voice and continuous wave net will be monitored by SFOB, C, and B detachment levels of command. This net will be used as an alternate means for subordinate detachments to contact higher headquarters when other means fail.

e. Equipment Augmentation. To accommodate traffic loads increased by counterinsurgency operations, communication equipment augmentation will be necessary. Provisions must be made for a command message net and command voice net at each echelon of command. In addition to the augmentation of personnel and equipment sent to the C detachment by the signal company, the C detachment will require radio equipment sufficient to operate a net control station for its command message net (radio-teletype), and a net control station for its command voice net. In addition it will require equipment to enter the

NET	SFOB	"C" DET	"B" DET	"A" DET
COMMAND MESSAGE NET (GP)	NCS	4		
COMMAND MESSAGE NET (C DET)		NCS	3	
COMMAND VOICE NET (GP)	NCS	4		
COMMAND VOICE NET (C DET)		NCS	3	
COMMAND VOICE NET (B DET)			NCS	4
EMERGENCY NET (GP)	NCS	4	12	48

Figure 54. Special Forces group counterinsurgency radio nets.

command voice net operated by the SFOB. The B detachment will require the same equipment augmentation. The A detachment will require equipment to enter the B detachment command voice net.

116. Wire

Long distance wire communications are not normally feasible in a counterinsurgency environment. Local wire systems in established Special Forces camps are highly desirable to provide—

a. Immediate contact with key personnel for radio wire integration calls.

b. Control and coordination of camp defenses.

c. Intra-camp administration and operations.

d. Communications with observation posts.

117. Training

a. The reorganization of the Special Forces group communications structure to accomplish a counterinsurgency mission requires that Special Forces personnel at all echelons be trained in the functioning of the signal communications systems.

b. Signal training also will be required for the paramilitary forces that are established under control of the Special Forces operational detachments. The Special Forces detachment radio operators are well qualified to establish tactical communications systems within company-sized, paramilitary units. Training will emphasize the following aspects of signal communications:

    (1) Radio installation.

    (2) Voice radio procedure.

    (3) Communications security.

    (4) Concept of communications.

c. The type of signal equipment provided to the paramilitary forces may be standard, obsolete, or current U.S. military equipment or civilian procured equipment designed for use in the particular area concerned. The equipment will be simple to operate and maintain.

d. Maintenance training beyond operator level should be conducted for selected indigenous personnel at a central location for the entire country or political subdivision.

118. Communications Security

a. The basic elements of communication security as outlined in paragraph 114 apply in counterinsurgency. Emphasis on certain aspects, however, will be different.

b. Physical security as applied in conventional units will apply to Special Forces operations in a counterinsurgency role.

    (1) Classified communications documents will be secured in compliance with AR 380–5, AR 380–40, and other classified communications publications.

    (2) Radio equipment will be provided the

normal security measures applicable to conventional units.

c. Transmission security will be emphasized in voice radio operations. Since the operations of the Special Forces detachments are not covert in nature, the need to avoid direction finding is not paramount. All possible steps will be taken to prevent traffic analysis.

d. Cryptographic security pertains in counterinsurgency operations.

CHAPTER 10

LOGISTICS

Section I. EXTERNAL LOGISTICS

119. General

Since logistical support for Special Forces conducting counterinsurgency operations is no different from that provided other U.S. Army forces, such support is not presented in this manual. In unconventional warfare, however, the sponsoring power provides supplies to the GWOA by delivering them to the SFOB. The SFOB, in turn, delivers these supplies to the operational detachment upon their request, and who, through the judicious use and issue of supplies, obtains the unity of effort and support required for U.S. objectives. See FM's 31-21, 31-21A, and FM 101-10-3 for additional information.

120. Considerations

Land, sea, or air transport is used to deliver supplies to the operational area. The requirements for secrecy during movement or delivery complicates such support and the detachment commander requesting extensive logistical support from the outside should hold his request to essential items not readily obtainable in the operational area. This could include major items such as weapons, ammunition, demolitions, communications equipment, medical supplies, or other items that are normally denied to the local population by the enemy. The detachment commander has several techniques available that will give him the supplies he needs when he needs them.

a. Accompanying Supplies. These are supplies taken into the operational area by the operational detachments during infiltration. These supplies are issued at the SFOB in the final briefing stages and are rigged by the detachment for delivery. The detachment commander considers the automatic resupply of survival and combat essential items that he will receive when he plans his accompanying supplies. These, plus the automatic resupply, will constitute a 30-day level of items needed for his operations. The supplies designated for the automatic resupply drop are selected on the basis of available intelligence indicating items essential to support the local resistance elements. Once drawn they are rigged and packed for drops and stored at the SFOB or at the departure site.

b. Automatic Resupply. The automatic resupply is prearranged for time, location, and content during the detachment's stay in the briefing center of the SFOB. The automatic resupply gives the detachment flexibility in that it may include back up communications equipment plus enough weapons, ammunition, demolitions, medicine, and other items to support training phases and small-scale, tactical operations.

c. On Call Supplies. After commitment into an operational area, and once communications have been established with SFOB, detachments are ready to begin requesting supplies based on operational needs and the capability of the indigenous force to receive and secure them. In order to expedite supply requests, insure accurate identification of needs, and minimize field station radio transmission time, the detachment commanders use the logistical brevity code system known as the Catalog Supply System (CSS) (app VII). The CSS is used to request three categories of supplies—combat essential items, such as weapons and ammunition; survival items such as medicines, clothing, food, and, the above listed items in bulk quantities to support rapid expansion of indigenous forces or greatly increased operations. CSS supplies are prepackaged in manportable, 50 pound bundles; equipped with pack boards and carrying straps; and packaged to resist damage on pack boards, from handling, storage, and weather. The code includes the general category, unit designation, unit weight, total bundle weight, and number of individual man-loads per package. Each load is self-contained. For example, a weapon will be packaged with ammunition, tools, cleaning equipment, and spare parts. Unused weight or space will be used for clothing, blankets, ponchos, or food and survival equipment.

d. Emergency Resupply. This resupply procedure is designed to restore the operational capability of the detachment after it has been lost because of enemy action, loss of scheduled radio contacts, or other incidents. Items delivered will consist primarily of communications equipment and other essential equipment necessary for survival.

121. Supply Accountability

Formal accountability ends when the supplies leave the departure point; however, the detachment commander is responsible for all supplies that enter the operational area. This is particularly true of sensitive items such as weapons, ammunition, demolitions, and other serial number items. This responsibility not only includes items given to the force by the sponsor but also those that have been put in the hands of the resistance force through battlefield recovery and capture.

Section II. INTERNAL LOGISTICS

122. General

Operational areas are expected to provide the bulk of food and clothing in the area. Support for indigenous personnel is affected by the three primary considerations—geographical location, the size of the force, and the type of operations. The geographical location will determine the type and extent of agriculture dominant in the area; to some extent it will also influence the diet of the local personnel. Geography influences the type and amount of personal clothing and equipment required and life expectancy of these items.

Diseases and noncombat injuries are greatly influenced by geographical location. The geography and enemy situation of an area determine the type of targets to be attacked which, in turn, influences the type of operations conducted. The size of the force to be supported is important. If local food procurement is adequate for the present force, the same may not hold true for larger forces developing later.

a. Replacement factors apply to items that break or wear out but do not lose their identity by use, e.g., boots, weapons, vehicles.

b. Consumption factors, on the other hand, apply to items that are physically consumed or change their identity by use. Examples are food, ammunition, demolitions, and medicine.

c. Available information enables the detachment commander to forecast needs and plan appropriate procurement well in advance. Reasonably accurate experience factors to guide the commander's planning can be developed if detailed issue records are maintained. For further information, see FM 31–21.

123. Civilian Support

In many cases civilian support of the guerrilla effort will be completely spontaneous; the defeat of a hated enemy being adequate payment. In other cases, it may be necessary to establish a barter system with cigarettes, medicine, drugs, or other high-value, low-bulk items being exchanged for goods and services. In some cases outright payment in currency, gold, or receipts for future payment may be demanded. The impact of increased currency flow in the local economy must be considered, as well as the security risk involved, before purchasing is adopted as a procurement system in a particular area. In some instances the area command may use the levy system. Each family or village is assessed its "fair share" based on ability to contribute. Payment, if any, will usually be deferred until after cessation of hostilities. Consideration must be given to the psychological impact on the local populace of the guerrilla force(s) living off the country.

124. Local Procurement

The area command is the focal point of the logistical effort. Each element of the area command—guerrilla force, auxiliary, and underground—is assigned a role in the logistical effort. The guerrilla force acts as a procurement agency in conjunction with operations against the opposing force. It also has the responsibility for most of the transportation required and for selecting and securing the caches used for day-to-day issue. The auxiliary force is the principal logistical element within the area command. Since they live at home and lead reasonably normal lives they are free to devote considerable time and effort to the procurement, movement, and storage of local supplies. The underground, on the other hand, should not be given extensive logistical responsibilities. Their area and methods of operation are such that the bulk of their effort is needed for self-support and for assigned clandestine missions. They can often provide vital support by procuring limited items normally denied civilians, such as drugs, chemicals for explosives, radio repair parts, and documentation.

125. Pre-Emergency Caches

a. It may be possible to establish pre-emergency caches. These caches are established in areas, presently under friendly control, which may become GWOA's during a major conflict. Stay behind operations lend themselves to this type of caching.

b. The major consideration for pre-emergency caches are (1) the probability that the cache will be needed, (2) the storage life of the item(s) involved; and (3) the problem of providing adequate security for the cache. Caches are not established haphazardly but are the result of requirements generated by specific operations plans (OPLANS).

126. Mission Support Site (MSS) Caches

Security will often appear to be at cross-purposes with supply planning. Security will dictate that caches be kept small, widely dispersed, and relatively inaccessible to the enemy. Consideration should be given to locating caches at planned or probable mission support sites (MSS). An MSS is a temporary base used by personnel who are away from their base camp, during an operation, for periods in excess of 2 days. The MSS may provide food, shelter, medical support, ammunition, or demolitions. The use of an MSS eliminates unnecessary movement of supplies and allows the indigenous force to move more rapidly to and from attack sites. When selecting an MSS, consideration is given to cover and concealment, proximity to the objective, proximity to supply sources, and the presence of enemy security forces in the area. Although transportation problems will be increased, security dictates that the DZ's and LZ's be a considerable distance from caches, MSS's, and base camps.

127. Decentralized System

As the indigenous force grows in size, subordinate units should be assigned a sector or zone of operations and be responsible for establishing a separate procurement system for their sectors. This will greatly reduce transportation needs since the supplies are procured near the consumer. This will also improve security since the compromise or destruction of the procurement system in one sector will not destroy the entire procurement apparatus. Movement of supplies between sectors is kept to a minimum; and names, storage sites, and caches are not passed from sector-to-sector. An additional advantage of a decentralized system is that it permits a more equitable distribution of the logistical burden on the civilian population. Detachment commanders considering decentralized supply systems, should consider air delivery direct to the user thus easing transportation requirements.

128. Transportation

a. Reception facilities at LZ, DZ, or water landing sites will determine the extent to which the detachment must depend on local transportation. Numerous and secure reception facilities permit external supplies to be delivered near the user. A single aircraft or other carrier can deliver portions of loads to two or more sites. This detachment would, in such case, submit separate resupply requests and DZ locations for each sector or operational area.

b. Local transportation available will vary from area-to-area. Transportation problems will increase as the detachments expand the size of the indigenous force and increase the scope of operations. Decentralized procurement and decentralized receipt of externally provided supplies will reduce this problem, but area commands should plan for a more formal transportation element during the more advanced phases of operations. Terrain and distance, in conjunction with transportation availability, are the primary considerations involved in selecting transportation to be used in a particular operation. The terrain will often be such that only animals or human bearers can be used. In other cases the distance and tonnages will be such that this type of transport is not capable of performing the mission efficiently.

c. Security, as applied to transportation, often revolves around the question of speed and payload versus secrecy. Off-road movement is not necessarily the most secure, since it means that men or animals must move for several days to perform the same task that one truck could perform in a few hours. Detailed intelligence concerning the enemy's off-road and on-road security systems will be invaluable in determining the mode of transportation to be used at any particular time.

d. Transportation will normally be provided by the auxiliaries on a mission basis since this permits them to continue to lead a "normal" life in the community. When the scope of operations dictates the need for a full-time transportation element the area command must program indigenous force personnel as vehicle drivers, animal handlers, or as bearers.

129. Maintenance

a. Operations will influence maintenance requirements. Weapons and other equipment used during active phases of operations are difficult to maintain. By careful planning, periods of relative troop inactivity can be used for individual maintenance.

b. Although the local maintenance skill levels will vary, operational areas are generally unsophisticated and people in such areas will be able to repair and improvise to a surprising degree. Detachment personnel must determine what local skills are available; enlist the people involved as supporters of the guerrilla movement; provide them with parts, tools, and kits as needed; and finally, provide limited technical training if needed. As operations progress, it may be desirable to establish centralized maintenance facilities, especially for such basic maintenance as repair of weapons, radios, boots, and clothing.

c. Maintenance training should receive high priority in a GWOA. Since comfort and survival depends upon proper maintenance of equipment, the individual guerrilla has motivation. His technical knowledge, however, may be limited. Radio operators, weapons crews, and operators of similar equipment must be carefully selected. Infiltration of maintenance technicians may be necessary. Emphasis, however, will remain on employment of local skills.

130. Construction

In GWOA's construction work will be held to a minimum because of security expenditure of time and energy, and the required immobilization of the guerrilla force; however, the exact opposite is required in support of a counterinsurgency operation (ch. 11).

a. First priority is basic medical facilities.

b. In areas of extreme temperature, personnel shelters will have priority.

c. In rainy areas storage facilities receive priority.

d. Classroom areas may be required in extremely cold climates.

e. Interference by enemy personnel can be minimized if construction work is conducted during periods of—
- (1) Extreme temperature.
- (2) High precipitation.
- (3) Diversionary small-unit operations relieving enemy pressure.

131. Labor

The bulk of the labor requirements in the GWOA will be met by the guerrilla force itself. When it is necessary to supplement this force it should be done on a mission basis, using members of the auxiliary as needed. Immediately before linkup it may be desirable to use all available personnel to assist in the construction of defensive works, road blocks, or other tactical construction.

132. Evacuation and Hospitalization

See chapter 12.

CHAPTER 11
DEMOLITIONS AND ENGINEERING

Section I. UNCONVENTIONAL WARFARE

133. Planning Considerations

When planning and preparing for commitment into operational areas to support an indigenous force in unconventional warfare, the Special Forces combat engineer specialist will conduct extensive training in the use of conventional and expedient demolitions. The primary consideration here is in the preparation and combat employment of a trained guerrilla force against an enemy. The engineer specialist will develop tactics in the use of destructive techniques in the interdiction of highways railroads, bridges, and other lines of communications. He will be prepared to train and assist selected auxiliary and underground elements in the construction and use of sabotage devices and other techniques designed to harass the enemy and cause him to divert his fighting force, as well as destroy his will to fight. In planning unconventional warfare operations, the detachment commander and his engineer specialist must consider the preparation of defensive positions, the construction of obstacles and boobytraps, and the use of antipersonnel and vehicular mines. The construction of small shelters for medical facilities and storage facilities for equipment and material may have priority in some areas. Climatic conditions will dictate the need for construction of living shelters and other facilities as well as field fortifications to protect outposts and defensive positions.

134. Destructive Techniques

Generally, the destructive techniques discussed here are applicable to unconventional warfare. In a counterinsurgency environment where Special Forces detachments are advising paramilitary and local indigenous forces in construction projects and establishment of defensive positions, the use of conventional demolitions may well be the only requirement. Conventional demolitions may be employed quite effectively in road building, land clearing, airlanding construction, and obstacle removal projects. Logistical procedures normally found in a counterinsurgency operation will permit the extensive resupply and use of conventional demolitions. However, when committed into an operational environment both unconventional warfare and counterinsurgency operations involving training and directing indigenous forces in combat operations, Special Forces personnel will find that the lack of logistical support, air delivery capability, and mobility may require using demolitions on a limited scale. The amount of demolitions that may be carried for operational use will result in Special Forces personnel improvising destructive charges and employing them to get maximum results from a minimum amount of material. FM 5–25 and FM 5–31 complement this section and should be used in conjunction with it.

135. Advanced Charges

These charges normally are fabricated from plastic explosives using principles of explosive force and direction to destroy or immobilize select targets.

a. Saddle Charge (fig. 55).

(1) *General.* The saddle charge is used for cutting steel bars and shafts 5.08–20.32 cm. in diameter. Turbine and propeller shafts, if motionless, are examples of targets on which saddle charges can be used. The saddle charge achieves results by employing what is known as the "cross fracture." The fracture forms below the base of this triangular shaped charge, cutting the steel target.

(2) *Preparation.* The saddle charge is shaped to form an isosceles triangle. The short axis (base) is one-half the circumference of the target. The long axis is twice the length on the base. The depth or thickness of the sliced plastic explosive is one-third a block of C4 for targets 15 centimeters and less in diameter; one-half a block of C4 if more than 15 centimeters and less than 20 centimeters for the hardest steel. Thick-

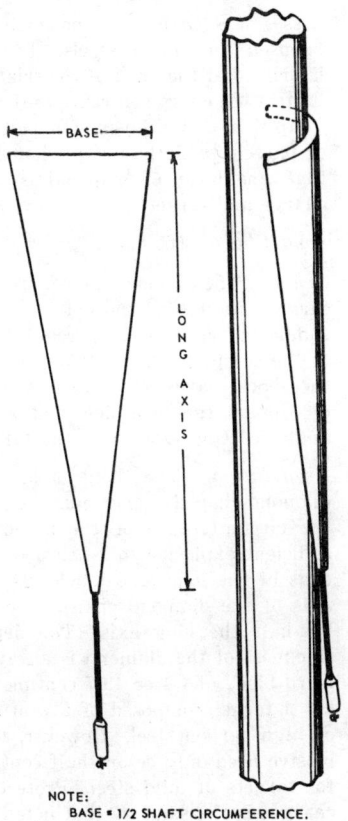

NOTE:
BASE = 1/2 SHAFT CIRCUMFERENCE.
LONG AXIS = 3 X BASE.

Figure 55. Saddle charge on a steel shaft.

nesses less than 2.54 centimeters may be used for milder steels. The charge is primed at the apex of the triangle. To protect the charge en route to the target, wrap it in a thin layer of paper, tinfoil, or parachute cloth; insure that no more than one layer of material is between charge and target.

b. Diamond Charge (fig. 56).

  (1) *General.* The diamond charge is used for targets similar to those for which the saddle charge is used but requires access to the complete shaft. When detonated, the shock waves, meeting in the center of the charge, are deflected at right angles cutting down into the target.

  (2) *Preparation.* The long axis of the diamond-shaped charge must be equal to the circumference of the target. Use sufficient explosive to be sure that both ends of the long axis touch. The short axis of the diamond charge is equal to one-half the long axis. The depth or thickness of the diamond is always one-third block of C4 or 1.67 centimeters to cut a target composed of 2 centimeters of high carbon steel. However, the explosive need only be one-half centimeter for targets of mild steel. Protection of charges is the same as that noted above for saddle charges. Simultaneous detonation is mandatory. Equal lengths

of detonating cord may be used with non-electric blasting caps crimped to the ends or electric blasting caps fired simultaneously.

c. *Counter Force.*
   (1) *General.* This charge will effectively breach dense concrete and occasionally certain timber targets up to 1.22 meters in thickness. Excellent results will be

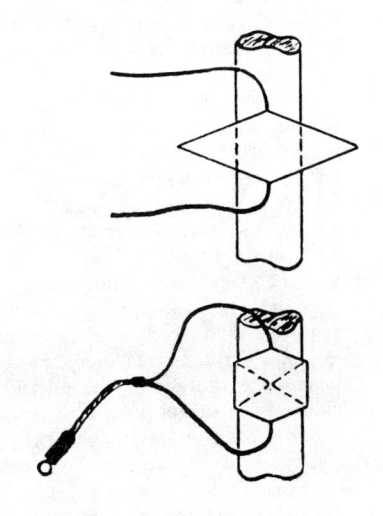

NOTE: CHARGE PLACED AGAINST SHAFT AND FORMED AS SHOWN.

Figure 56. Diamond charge.

obtained with relatively small amounts of explosive when properly constructed, placed, and activated. The simultaneous activation of two diametrically opposed charges on the target causes the shock waves to meet at the target center. The resultant pressure causes internal damage.

(2) *Preparation and placement.* The actual size of the charge is governed by the target thickness, in meters, of the target to be breached. Multiply the diameter or thickness of the target to be breached by the constant, 5, which gives the number of pounds of plastic explosive required for reinforced concrete. Round off to the next higher meter for any fractions less than 1 meter. For example, if a concrete target pier measures 1.06 meters in thickness, the total amount of plastic explosive required is 6 pounds (2.72 kilograms). Divide the required amount of explosives in half and place the halves diametrically opposite one another on the target. Both sides of the target must be accessible in order to position the two charges against the sides of the target. To secure the charges against the target one of the following methods may be used:

(*a*) Suspend the charges from ropes which pass over the top of the target (fig. 57).

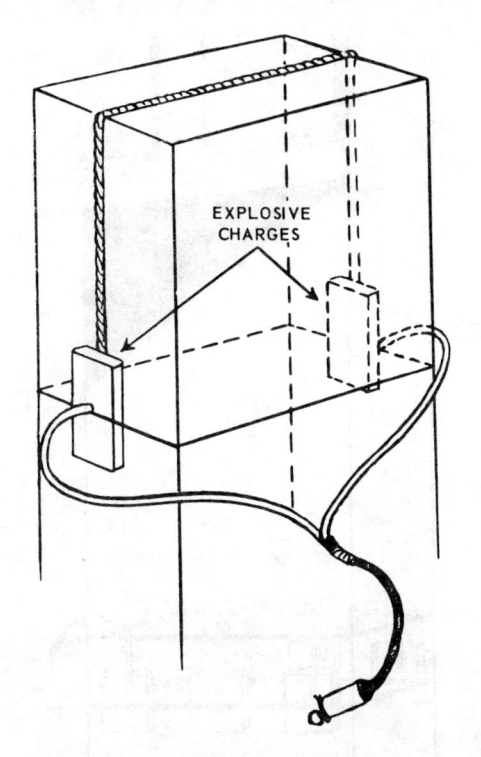

Figure 57. Opposed charges suspended by rope.

    (*b*) Construct a simple frame of a size that will enclose the target. Secure the charges to both sides of the frame so that they are diametrically op-

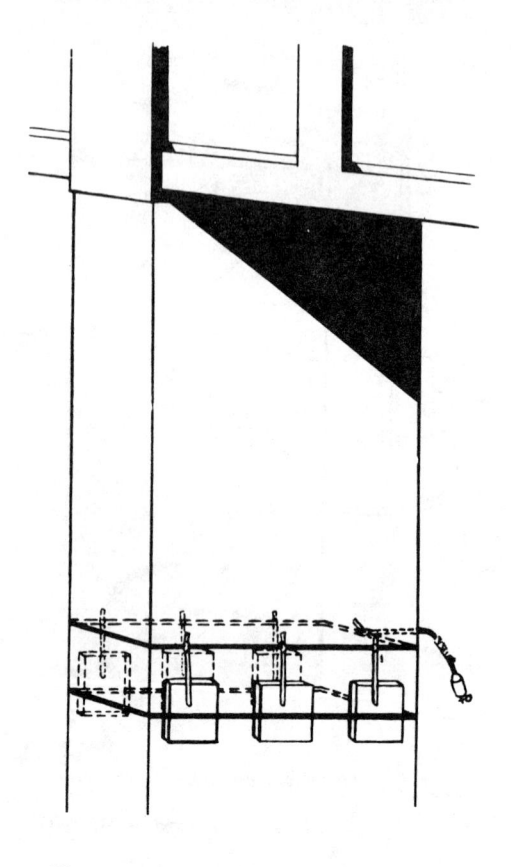

Figure 58. Opposed charge placed on bridge pier.

posed. To secure the frame in position one side may be hinged and the frame wedged in place (fig. 58). The frame may also be suspended using ropes which pass over the top of the target.

(c) On relatively small targets, visual adjustment and placement may be employed. Charges are held in place by propping them with boards or poles.

(3) *Priming.* Simultaneous detonation of both charges is mandatory. The most common procedure is to crimp nonelectric blasting caps to equal lengths of detonating cord; prime at the center rear of the charge; join the two free ends together at a point 15.24 centimeters from the end, and tightly tape the blasting cap of the firing system to the parallel detonating cords.

d. *Platter Charge* (fig. 59).

(1) *General.* The platter charge is most effective against POL storage containers, transformers, and similar thin-skin targets that are usually made inaccessible by fencing. The size of the charge is governed by the weight of the platter. The configuration of the platter may be any shape; but, concave shape raises the temperature of the platter during flight, thereby assisting in igniting the POL upon hitting the target. The explosive, upon detonation, projects the platter

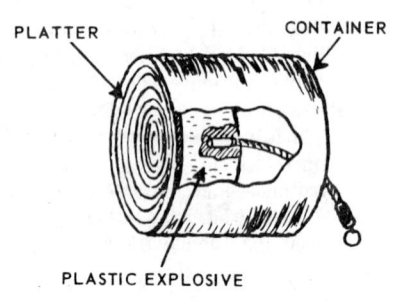

Figure 59. Platter charge.

through the air and into the target. If a chain-link (cyclone) fence is between the charge and the target, the fence will be penetrated and the platter will continue to the target.

(2) *Preparation of charge.* A container of any kind with a diameter which roughly coincides with the diameter of the platter may be used. Both ends of the container should be removed. Position the platter at one end of the container with the concave side facing out. A platter of any material other than metal will result in a greatly reduced range. Steel provides the best result with effective ranges of more than 40 meters. The amount of explosive required to propel the platter should equal the weight of the platter. The explosive is then packed firmly in the container.

When no container is available, the explosive may be taped to the platter and hung by coat hangers on fence.

(3) *Placement of charge.* The platter container is positioned on its side with the concave face of the platter directed toward the target. The maximum, effective range is dependent upon the size of the target, and may extend to 40 meters or more.

(4) *Priming.* Prime the charge exactly in the rear center. Exact rear-center detonation of the charge is essential for uniform distribution of shock waves and proper propulsion.

e. Ribbon Charge (fig. 60). This charge, if properly calculated and placed, cuts mild steel up to 5 centimeters in thickness with considerably less explosive than formula computed ($P = 3/8A$) charges. It is effective on noncircular steel targets up to 5 centimeters. It can be shaped for use against I- and T-beams. On the corners and ends where the detonators are placed, it may be necessary to "build up" this area with additional explosives since the charge will be less than 1.27 centimeters thick.

(1) Thickness of charge = the thickness of the metal.

(2) Width of charge = twice the thickness of the charge.

(3) Length of charge = the length of the cut.

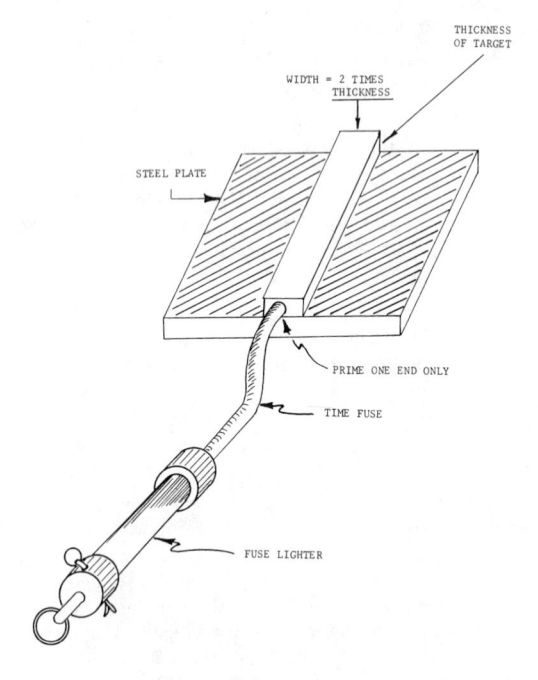

Figure 60. Ribbon charge.

136. Improvised Charges

These are charges normally made from standard explosive materials such as plastics explosives, but using improvised casings and firing devices and attaching additional metal for a more destructive effect.

a. Shaped Charge.

   (1) *General.* A shaped charged is designed to concentrate the energy of the explosion on a small area to make a tubular or linear fracture in the material on which it is placed. The versatility and

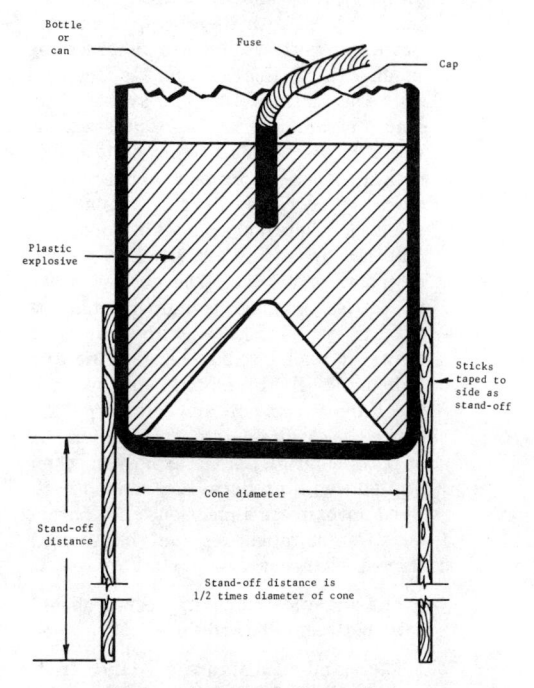

Figure 61. Improvised shaped charge.

simplicity of shaped charges make them an effective weapon, especially against armor plate and concrete. Shaped charges are easily improvised.

(2) *Improvised shaped charges* (fig. 61). Almost any conically shaped container may be used to make a shaped charge; however, best results are obtained by using a cavity liner of 3-mm copper, and with the angle of the cone at 42°. Tin, zinc, or cadmium are also satisfactory. Cups, bowls, funnels, wine bottles, and cocktail glasses are items which may be used. If no cavity liner is available, a reduced, shaped-charge effect may be obtained by cutting a cavity into a block of plastic explosive. The narrow necks of bottles or glass containers may be cut by wrapping them with a piece of twine or string soaked in gasoline and lighting it. Hold the bottle upright until the flame goes out and submerge the bottle, neck first, into cold water. A narrow band of plastic explosive, when ignited, will produce the same effect. The approximate dimensions and characteristics suitable for an improvised shaped charge are—

(a) The angle of the cavity (cone) should be between 30° and 60°.

(b) The stand-off distance (distance from the bottom of the shaped charge to the target) should be from 1 to 2 times the diameter of the cone.

(c) The height of the explosive content, measured from the base of the cone, should be twice the height of the cone.

(d) The detonation point should be exact top center.

(e) An ogive must be used if charge is placed underwater.

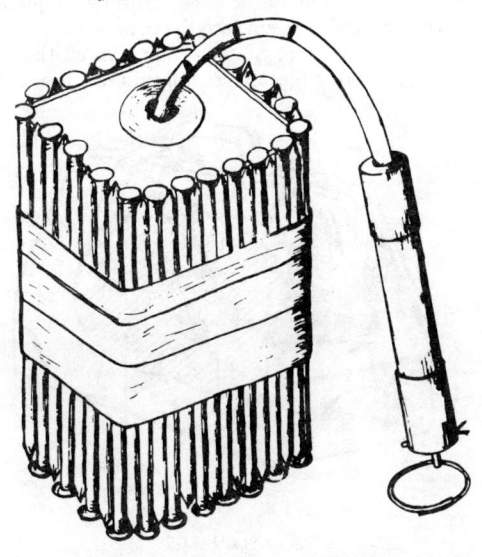

Figure 62. Improvised grenade.

b. Grenades (fig. 62). A simple grenade can be constructed by the use of a 0.225-gram block of explosive, scrap metal, time fuse, and non-electric cap. The time fuse is used as the delay element. Bolts, nuts, nails, or other pieces of metal are secured to the grenade for fragmentation effect. A small metal pipe may also be used as the grenade jacket. The grenade is detonated by a nonelectric blasting cap crimped to a short piece of time fuse. Insert the blasting cap into the explosive and tie or tape it firmly in place. Small V-notches are cut into the fuse. As the time fuse burns, a spurt of flame appears at the V-notches. After the flame appears at the last

COVER CHARGE

MAIN CHARGE

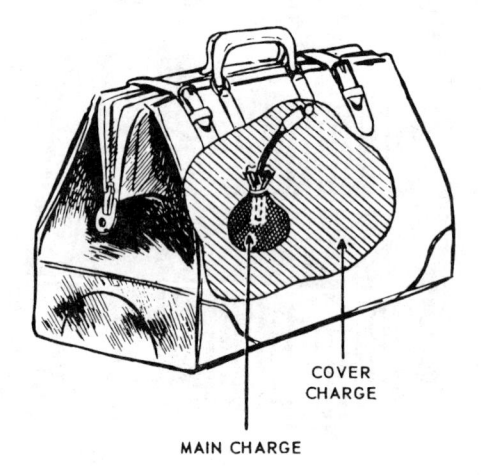

Figure 63. Dust initiator.

V-notch, the grenade is thrown. If desired, only one V-notch (closest to the cap end) may be used. The V-notches should be taped to keep out moisture.

c. Dust Initiator (fig. 63). The dust initiator is used to destroy buildings and certain storage facilities. The basic purpose of this device is to achieve two distinct but rapidly successive explosions. This is accomplished by constructing a main charge composed of equal amounts of powdered TNT (obtained by crushing TNT in a canvas bag) and magnesium powder and inclosing it in a cover charge of a scattering agent of any carbonaceous material which can be reduced to dust vapor. Examples are cornstarch, flour, coal dust, or gasoline (when using gasoline, never use more than 11.4 liters.) Thermite may be substituted for magnesium in the mixture. From 1.36 to 2.67 kilograms of surround should be provided for each 28.32 cubic meters of target. The 0.45 kilograms (0.225 kilograms TNT and 0.225 kilograms magnesium powder) charge will effectively disperse and detonate up to 18.14 kilograms of carbonaceous material. Upon activation, the main charge detonation distributes the cover charge material, which is initiated by the action of the incendiary explosion. This causes the entire atmosphere to be saturated with burning materials. The destructive effect of this device is increased by closing all windows and doors in the target building. This charge may be detonated electrically or nonelectrically.

Table I. Dust Initiator, Size of Building vs. Amount and Type of Cover Charge.

Target size	Cover charge	Amount	Type target
4,300 cubic feet (123 cubic meters)	Wheat flour	10 pounds (4.5 kg.)	Wooden building
4,300 cubic feet (123 cubic meters)	Coal dust	10 pounds (4.5 kg.)	Wooden building
18,500 cubic feet (524 cubic meters)	Wheat flour	100 pounds (46 kg.)	Wooden building
450 cubic feet (13 cubic meters)	Gasoline	6 gallons (23 liters)	Room
159,000 cubic feet (4,503 cubic meters)	Gasoline	30 gallons (115 liters)	Building
88,000 cubic feet (2,492 cubic meters)	Gasoline	2½ gallons (10 liters)	Cold storage room

*Extracts from technical report 2448 "T15 High Explosive Dust Initiator (U)" dated August 1957. Samual Feltman Ammunition Laboratories, Picatinny Arsenal, Dover, N.J.

d. Use of Explosives in an Ambush (figs. 64 and 65).

(1) Improvised and issued grenades, mines, mortar and artillery rounds, bangalore torpedos, etc., may be used in ambushes. This example uses grenades, detonating cord, and firing devices. To prime the grenade, unscrew and remove the fuze, crimp a nonelectric blasting cap to a detonating cord branch line, and insert the blasting cap into the fuze well. Attach the grenades, at varying heights, on trees bordering the avenue of approach and camouflage them. Link and camouflage the detonating cord branch line to a concealed main line or preferably a ring main. Attach a camouflaged trip-wire or pressure device if it is desired that the enemy activate the system. These charges may also be activated by a concealed guerrilla using an electric or nonelectric firing system. Trip-wires attached directly to loosened safety pin rings of grenades may also be used as an ambush weapon.

(2) Another excellent anti-personnel device for ambushes is a fougasse. This device is shown in figure 65.

e. Claymore (fig. 66). The improvised claymore approximates the fragmentation effects of the claymore anti-personnel mine (M18 or M18A1). To construct an improvised claymore,

Figure 64. Hasty ambush.

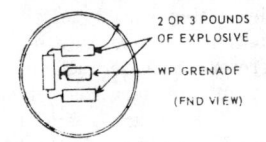

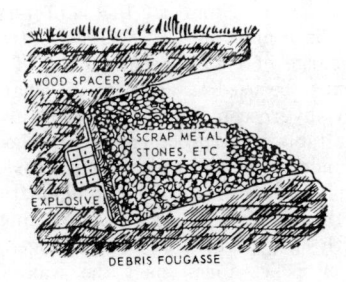

Note. Fougasses may be fired electrically or nonelectrically. Lightweight camouflage should cover exposed portion.

Figure 65. Types of fougasses.

use a container similar to a 3.785-liter can (#10). Place assorted metal fragments in the bottom of the container and cover with a buffer made of cardboard, leaves, grass, wood, or felt disk. Press plastic explosive firmly on top of the disk. The explosive-fragment ratio by weight, is 1 part explosive to 4 parts of metal. The weapon is positioned and aimed similar to a rocket fired without the launcher. For best results, prime the charge at the exact center point.

f. Ammonium Nitrate Fertilizer (AN). An explosive with a detonating velocity of three to four thousand meters per second may be made by using oil, gas, kerosene, or diesel fuel, and $33\frac{1}{3}$ percent ammonium nitrate fertilizer. Pour 2 liters of fuel into a 23-kilogram bag of prilled (small pellets), ammonium nitrate fertilizer and allow it to stand for at least an hour so that the pellets absorb the oil. Number 2 fuel is preferred and the fertilizer must be $33\frac{1}{3}$ percent nitrate. A booster charge of 0.45-kilograms of TNT, or equivalent must be used to detonate the charge. AN is very hygroscopic (absorbs moisture); therefore, suitable waterproof containers should be used for underwater and for prolonged periods of underground emplacement prior to detonation. Another method is the use of wax. Utilizing a fertilizer with $33\frac{1}{3}$ percent or more nitrogen in the prilled or pellet form, melt the wax and slowly add the fertilizer while stirring. A container (#10 can) or sack is then filled with the mixture of 0.225-kilogram block of TNT is added before the wax cools. The TNT acts as a booster,

- PROJECTILES
- WOOD SPACER
- EXPLOSIVE

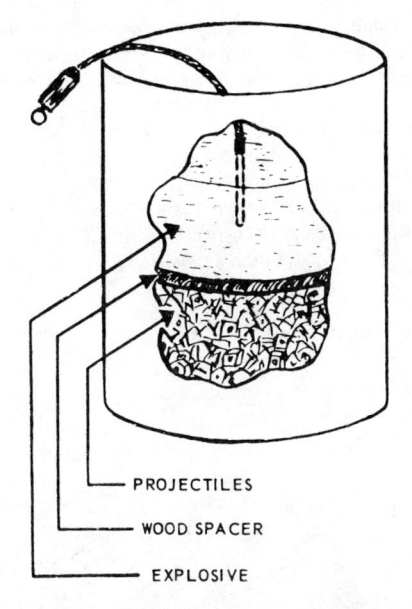

Figure 66. Improvised claymore.

and handles may be added to simplify carrying. This charge may be stored for a considerable length of time without a noticeable difference in strength.

137. Expedient Use of Standard Items

a. Firing Rockets Without a Launcher. When rocket launchers are not available, rockets may be fired using improvised techniques. The rocket must be placed at least 4 meters away from the target to permit discarding of the bore-riding safety pin, the rocket may be armed. Any type of launcher, such as a V-shaped trough or a pipe, may be used but should be at least twice as long as the rocket for ranges above 45 meters. Launchers no longer than the rocket may be used for ranges less than 45 meters (fig. 67). A salvo of rockets, fired electrically, provides excellent area coverage from a defensive position.

b. Electric Firing (fig. 68).

   (1) Disregard all but the two, white, plastic covered wires.

   (2) Strip the plastic coverings to expose the bare wire.

   (3) Connect the bared wires to the firing wire using a single twist.

   (4) Remove the bore-riding safety band and place the rocket in the firing platform so that the bore-riding safety pin is depressed.

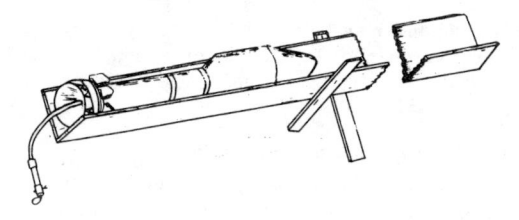

Figure 67. Expedient launching platform.

(5) Remove the shorting clip.

(6) Aim the rocket.

(7) Attach an electrical source and fire.

c. Nonelectric Firing. (fig. 69).

(1) Remove the wires from the fin assembly.

(2) Remove the plastic plug (cone) from the opening in the rear of the rocket venturi nozzle by prying it out with a nonsparking tool.

(3) Cut the end of a piece of time fuse and insert a match head. Tape matches around the fuse with the match heads directly over the end of the fuse.

(4) Drop 6 to 12 match heads into the rear of the rocket, ensuring that they are next to the propellant sticks.

(5) Remove the bore-riding safety band and place the rocket as described in *b*(4) above.

(6) Place the time fuse with match heads snugly against the disc-perforated separator.

(7) Aim the rocket and light the time fuse. The time fuse ignites the matches which in turn ignite the rocket propellant.

d. Aiming the Rocket. The cardboard rocket container may be used as an aiming tube. Place string across the open end of the tube to form a cross (fig. 70). Sight through the aperture of tube and adjust the firing platform until the de-

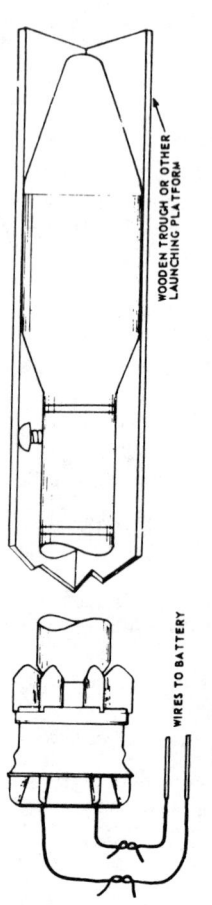

Figure 68. Electrical firing of 3.5-inch rocket without launcher.

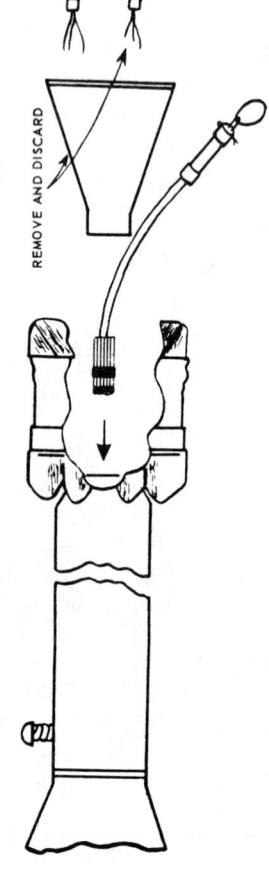

Figure 69. Nonelectric procedure for firing a 3.5-inch rocket.

50 METERS
100 METERS
200 METERS

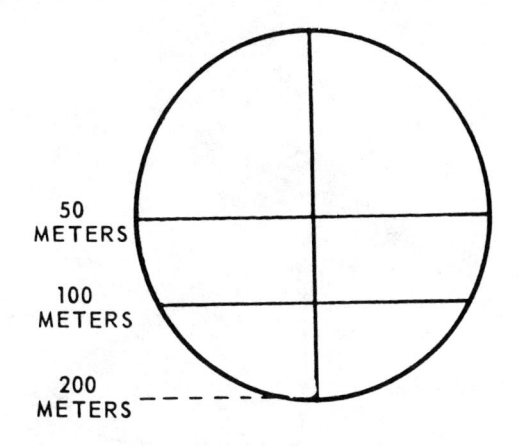

Figure 70. Crosshair at open end of rocket tube.

sired sight picture is obtained. Remove the tube and place the rocket on the firing platform. The metal shipping container may be used as a firing platform.

e. Using the Rocket Head as a Shaped Charge. The rocket head may be separated from the fuse-motor assembly and used as a stationary shaped charge.

   (1) Grasp the rocket head with one pipe wrench and the fuse and motor assembly with another (fig. 71). (Caution—carefully remove the rocket motor assembly; a red dot on the fuse is stab-sensitive.)

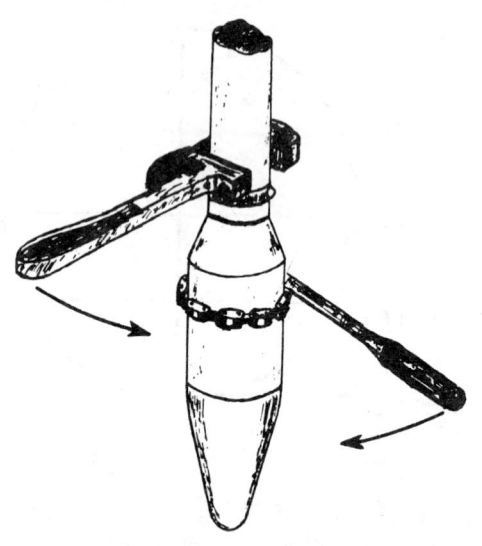

Figure 71. Removal of rocket head with wrenches.

    (2) Unscrew the motor-fuse assembly from the rocket head.

    (3) To prime the rocket head, use plastic explosives with a triple roll-detonating cord knot or a blasting cap (fig. 72). Secure the primer to the rocket head with tape.

    (4) Tape sticks to the rocket to maintain rocket head in an upright position. Proper stand-off distance is provided by the rocket head (ogive). Cut the ship-

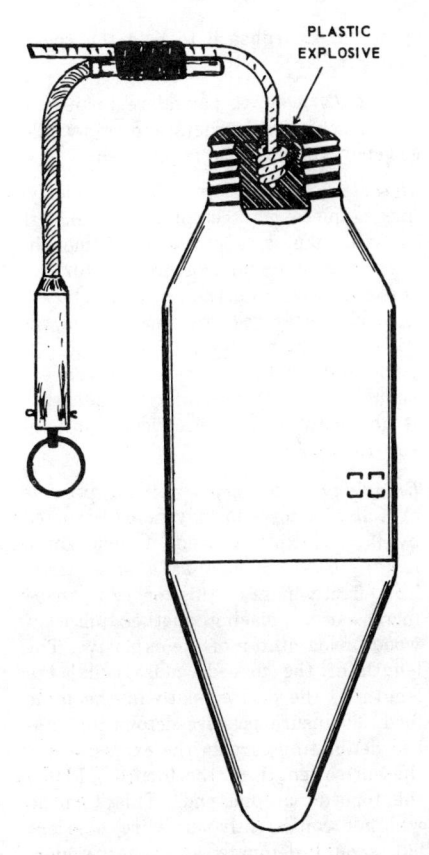

Figure 72. 3.5-inch rocket head nonelectrically primed with detonating cord.

ping tube cardboard to hold the rocket upright.

f. Bangalore Torpedo. A bangalore torpedo is designed to breach wire barriers and minefields. In an emergency, they may be improvised.

(1) *Breaching barbed wire.* Tamp explosive into a piece of steel pipe long enough to span the wire obstacle. Prime the torpedo at one end. Position the torpedo under the wire on the ground. The explosion causes the fragments from the pipe to cut the wire, thus creating a path through the barrier. This torpedo should also detonate mines underneath it and may be fired electrically or nonelectrically.

(2) *Breaching mine fields.* If no pipe is available, a torpedo may be constructed by taping explosives end to end on a length of wood such as a small tree. The effectiveness of this torpedo can be increased by placing another piece of wood or log on top of the explosive. The length of the torpedo must equal the length of the desired path in the mine field. To insure positive detonation, imbed detonating cord in the explosive for the entire length of the torpedo. Prime the torpedo at one end. This torpedo will not consistently cut wire barriers but should detonate all mines underneath it.

138. Improvised Incendiaries

a. General. Improvised incendiaries may be used when—

(1) It is desirable for security reasons to use an item manufactured within the operational area.

(2) Logistical difficulties prevent the delivery of needed manufactured items to the desired area of employment.

b. Chemical Mixtures. This section discusses chemical mixtures used to fabricate incendiaries of various kinds. Some of the desirable characteristics of incendiaries are—

(1) Easy to ignite.
(2) Difficult to extinguish.
(3) Burn with an intense heat.
(4) Leave little or no evidence.

c. Chemical Formulas. The names of certain chemicals and compounds do not always translate exactly into a foreign language. To eliminate this difficulty, table II, chemicals used internationally, shows alphabetical and numerical abbreviations for each chemical. The name of the chemical can be determined by personnel having a knowledge of chemistry using these abbreviations and reference materials such as encyclopedias and dictionaries.

d. Precautions. The following precautions should be observed when making improvised incendiaries:

(1) Use a mixing container made of non-sparking material such as ceramic bowl, cardboard, or newspaper.

(2) Use a wooden stick, plastic, or rubber spatula to stir chemicals.

(3) Do not heat gasoline or any petroleum product over an open flame.

(4) Do not place a top on the container when heating gasoline or wax.

(5) The vapors and dusts of many chemicals are toxic; therefore, they should be prepared in the open, or, if indoors, good ventilation should be provided.

(6) Test all finished products before using against a target.

(7) Store them in a dry place.

139. First-Fire Mixtures

a. Sugar-Potassium Chlorate Incendiary. A fast-burning, easy-to-ignite incendiary may be made by mixing 3 parts potassium chlorate or sodium chlorate with 1 part common household sugar. This mixture may be ignited by applying heat, spark, or sulphuric acid. It may be used as a primer (first-fire mixture) to ignite other mixtures.

b. Sugar-Potassium Permanganate Incendiary. Mix 1 part sugar with 9 parts potassium permanganate. It may be ignited by glycerine, time fuse, or spark.

c. Potassium Nitrate-Sulphur Incendiary. Mix

7 parts potassium nitrate (saltpeter) with 1 part sulphur and 2 parts flour, starch, coal dust, or sawdust. This may be ignited by flame or time fuse. Either sodium nitrate or ammonium nitrate may be substituted for potassium nitrate.

d. Potassium Permanganate-Aluminum Incendiary. Mix 2 parts potassium permanganate with 1 part aluminum. This mixture should be ignited with a time fuse.

e. Powder-Aluminum Incendiary. A very hot incendiary may be made by mixing 1 part black powder with 1 part aluminum. Either black powder or smokeless powder may be used. Smokeless powder may be obtained by pulling the bullets out of cartridges and pouring the powder from them.

140. Main-Fire Mixtures

The following mixtures of the same quantity will burn longer than the mixture discussed above. Main-burning mixtures are usually primed by one of the incendiaries discussed above.

a. Gelatin Gas. Use 4 parts nondetergent soap to 6 parts gasoline, kerosene, or other petroleum products. Heat the liquid in a double boiler or over a flameless heat source until it begins to boil. Then remove the liquid from the heat and introduce the soap in small chips or powder. Stir this mixture until it becomes a thick putty-like mass. This incendiary may be ignited with any flame.

b. Wax and Sawdust. Mix 5 parts wax and 5 parts sawdust. Any flame may be used to light this mixture.

141. Improvisation

a. Improvised Thermite Grenade (fig. 73). If issued thermite grenades are not available, one may be improvised in the following manner:

  (1) Pour about one-half centimeter of magnesium into a ceramic or clay container. This container must have a hole in the bottom covered with paper.

  (2) Add a mixture of 3 parts ferric oxide and 2 parts aluminum powder over the magnesium.

  (3) Add a priming mixture such as 1 part sugar and 3 parts potassium chlorate.

  (4) The priming mixture is ignited by using flame or sulphuric acid. When this grenade is lighted, the hot molten iron and aluminum pours out of the hole in the bottom of the container, burning through the target or welding parts together. A substitute for the ferric oxide and aluminum mixture is 1 part thermite with 4 parts magnesium.

b. Brick incendiary. An incendiary may be made to look like a building brick. Use 1 part water, 1 part plaster of paris, and 1 part powdered aluminum. The amount of each material must be calculated by weight. Mix the plaster

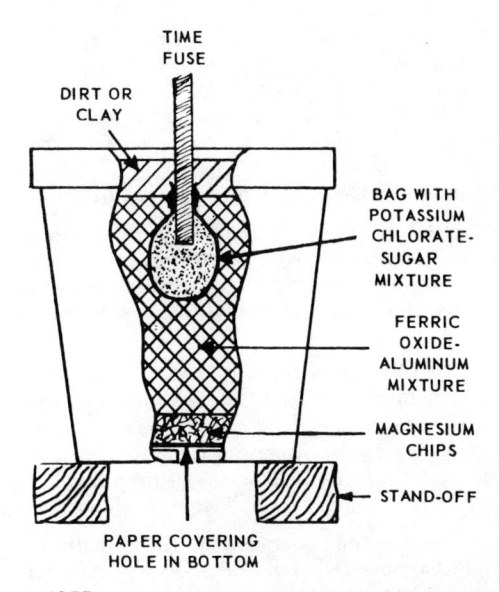

NOTE:

While a flower pot is shown as the container, a metal can may also be used. Five or six 1/4" holes are punched around the side at the bottom of the can. Cover these holes with one layer of cellulose or paper tape. No standoff is required since burning materials escape through holes on side of can.

Figure 73. Improvised thermite grenade.

of paris and aluminum. Add the water and stir rapidly. Make a hole for a primer before the brick hardens. The primer may consist of magnesium, or thermite. This incendiary may be made the color of brown or red brick by adding brick dust or iron oxide (Fe'''), depending upon the color desired. A cardboard or wooden form may be used to obtain the proper configuration.

142. Incendiary Mixtures and Igniters and Delays

a. General. This paragraph contains examples of easily constructed igniters and incendiary mixtures. A low-order explosion may be obtained by placing some of these mixtures in containers and detonating.

b. Cigarette Delay (fig. 74). Inclose the lighted cigarette in the matchbook or box and surround it with inflammable material such as rags, waste, or shredded paper. American cigarettes burn at the rate of 2.54 centimeters per 7 or 8 minutes in the air.

c. Candle Delay. Surround the candle with inflammable material, such as rags, wastes, or shredded paper.

d. String fuse. If time fuse is not available, it may be improvised as follows:

(1) Wash a shoelace or string in hot soapy water to remove the oil and dirt and rinse it in fresh water.

(2) Dissolve 1 part potassium nitrate or potassium chlorate and 1 part granulated sugar in 2 parts hot water.

(3) Soak the string in the hot solution for at least 5 minutes.

Table II. Chemical Formulas.

Chemical	Formula	Normal source
Potassium chlorate	$KClO_3$	Drug store*, hospital, swimming pool, and gymnasium.
Potassium permanganate	$KMnO_4$	Same as $KClO_3$.
Potassium nitrate	KNO_3	Fertilizer manufacturer and explosive plants.
Sodium nitrate	$NaNO_3$	Fertilizer manufacturer and glass foundries.
Ammonium nitrate	$(NH_4)NO_3$	Same as KNO_3.
Ferric oxide	Fe_2O_3	Same as KNO_3.
Powdered aluminum	Al	Electrical equipment, auto and paint manufacturer, paint store.
Magnesium	Mg	Auto manufacturer, machine shop, commercial chemical house.
Glycerine	$C_3H_5(OH)_3$	Drug store, soap and/or candle manufacturer.
Sulphuric acid	H_2SO_4	Garage, machine shop, school or hospital.
Sodium chlorate	$NaClO_3$	Match manufacturer and explosive plants.
Sulphur		Drug store, match manufacturer.

*Apothecary, pharmacy, chemical wholesale or retail shop.

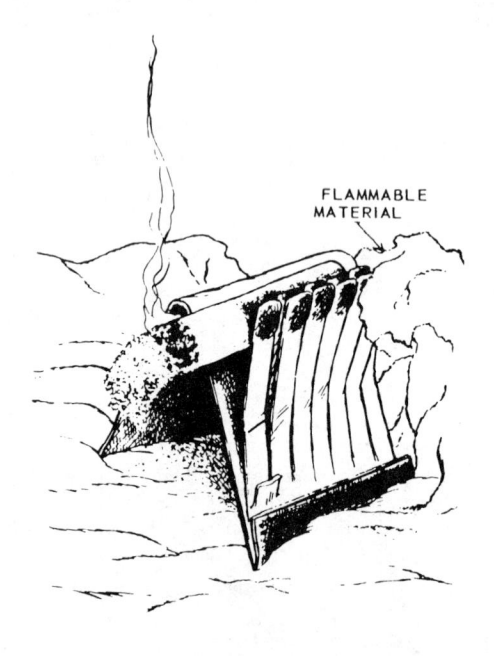

Figure 74. Cigarette delay.

(4) Remove the string from the solution and twist or braid three strands together and permit it to dry.

(5) Check burning rate by measuring the time it takes for a known length to burn.

e. Acid Delay. Acid delays may be constructed in various ways depending upon the material

available. The pipe incendiary (fig. 75) delay is
one example and is constructed as follows:

(1) Place a tight-fitting copper disk midway in a pipe.
(2) Fill one end of the pipe with a mixture of 3 parts potassium chlorate and 1 part sugar, then cork.
(3) Fill the other end with sulphuric acid and cork. When the pipe is placed with the acid higher than the sugar chlorate mixture, it slowly dissolves the copper disk, ultimately reaching the sugar chlorate mixture. This mixture produces a hot flame. The thickness of the copper disk, strength of the acid, and the temperature determine the length of delay.
(4) Gelatin capsules, rubber containers, or bottles with rubber membranes are other examples of materials that may be used with acids to achieve delays.

f. Water Can Delay (fig. 76) This device is improvised as follows:

(1) *Materials needed.*
  (a) A bucket-type container.
  (b) A float (wood or cork).
  (c) Small diameter floatmast.
  (d) Battery.
  (e) Electric blasting cap.
  (f) Electric wire.
(2) *Directions.* Make a small hole in the container. Attach the mast to the float. Place a copper wire through the diam-

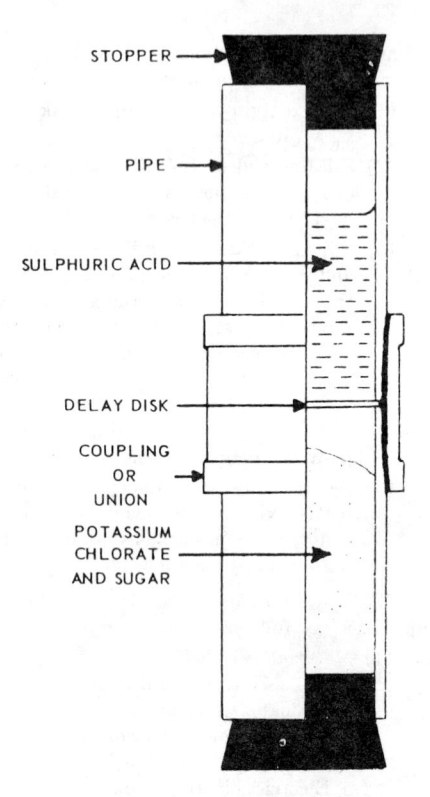

Figure 75. Incendiary delay in pipe.

eter of the upper part of the container with the insulation removed at the center. Prime the charge with an electric cap attaching one lead wire to the stripped end of the wire in the upper portion of the container, and the other lead wire to one of the terminals on the battery. Fill the container with water. Connect a wire from the other battery terminal to the top of the mast.

(3) *Functioning.* As the water drips from the container the float sinks in the can. When the top of the mast contacts the naked cross wire, the electric circuit is completed thus detonating the cap. Delay depends on the quantity of water and size of the escape hole. Some protection should be used to prevent falling trash, leaves, and other materials from stopping up the hole.

g. Watch Delay (fig. 77). This device is improvised as follows:

(1) *Materials needed—*
  (a) Watch with celluloid crystal.
  (b) Small screw (preferably brass or copper).
  (c) Battery.
  (d) Electric blasting cap.
  (e) Electric wire.

(2) *Directions.* Drill a small hole one-half centimeter from the center of the crystal and insert a screw. Tighten the screw so that either the hour or minute hand of the watch will make contact but the screw does not touch the face of

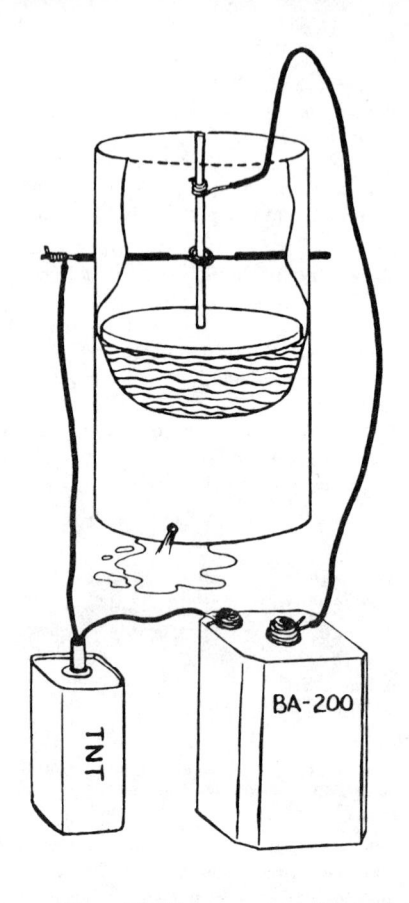

Figure 76. Water can delay (electric).

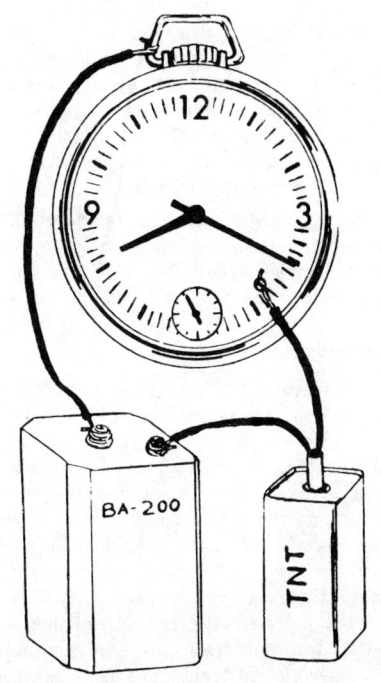

NOTE:
CAREFULLY REMOVE ANY FINISH FROM PORTION OF WATCH HAND TO TOUCH SCREW THRU CRYSTAL TO INSURE GOOD CONTACT. SHORT PIECES OF CAP LEAD WIRE MAY BE USED TO MAKE ILLUSTRATED CONNECTIONS.

Figure 77. Watch delay device (electric).

the watch. If a delay of more than 1 hour is desired, remove the minute hand. Wind the watch and set the hand for the desired delay. Connect one wire to the stem of the watch and a terminal of the battery and the other wire to the screw in the crystal of the watch.

(3) *Functioning.* When the hand of the watch comes in contact with the screw, the electric circuit is completed thus detonating the electrical cap.

143. Train Derailment

A number of factors influence train derailment. Security measures, such as patrols and trackwalkers, may be expected wherever there is resistance activity. Mountainous terrain offers steep grades, sharp curves, bridges, culverts, and tunnels which are ideal locations for derailment. In other areas a long level section of track may be the only vulnerable point, and it is suitable if the train moves through the area at high speed. Derailment on double-track lines should be accomplished on curves so that a single train will obstruct both tracks. Three derailment techniques are described in this section.

a. Technique, 3–5–2 (fig. 78). This method uses three charges, weighing 1.25 kilograms each, placed and tamped under every fifth crosstie.

(1) The charges are linked together with detonating cord. A firing device with detonator is taped to detonating cord

leads on both ends of the chain. The charges may be detonated from either end by an electrical firing circuit or by a pressure device actuated by the weight of the train wheels. The detonating cord should be extended on each end of the charges for a distance of at least 10 meters. The firing device should be placed at the end of the detonating cord nearest the approaching train. This is to insure that the charges blow in front of the train and not under it. The charges are placed under their respective ties and firmly tamped into position. The explosion will remove the rails at least 1 meter beyond the outside charges, disrupting at least 6 meters of rail. Additionally, the ties are broken and a crater is formed. The depth of the crater depends on the type of ballast material in the railbed. Since the charges are not placed in contact with the rail, the rail is lifted upon detonation. The pressure breaks the rail just beyond the outside charges.

(2) The disadvantages of this technique are the time required and noise associated with the placement of the charges. Once the charges are emplaced, they may be left in position for extended periods.

b. Technique, 10–2–1 (fig. 79). This method uses a total of ten 0.5 kilogram charges, each charge placed against the rail over every second crosstie.

> (1) The charges are linked together with detonating cord. The charges are wedged or lashed to the web of the rail directly over the crossties. Detonating cord priming leads must extend from both ends of the chain. A pressure firing device may be used.
>
> (2) The advantages of this technique are speed and silence in emplacement. The disadvantage is that the charges are visible to trackwalkers.

144. Foreign Explosives

Foreign explosives and equipment should be used when available. Except for minor differences, foreign material is similar to American.

a. Principal Explosives. Many countries make TNT, dynamite, and plastic explosives similar to American explosives. Table III shows standard explosives. It does not indicate the packaged size or form of the different explosives.

b. Characteristics of Foreign Explosives. A particular explosive produced by one nation is usually similar in characteristics to the same explosive produced in another nation. Minor differences in purity, density, ingredients, etc., may influence the performance of an explosive slightly; but the important characteristics particularly those of sensitivity and stability, are generally the same.

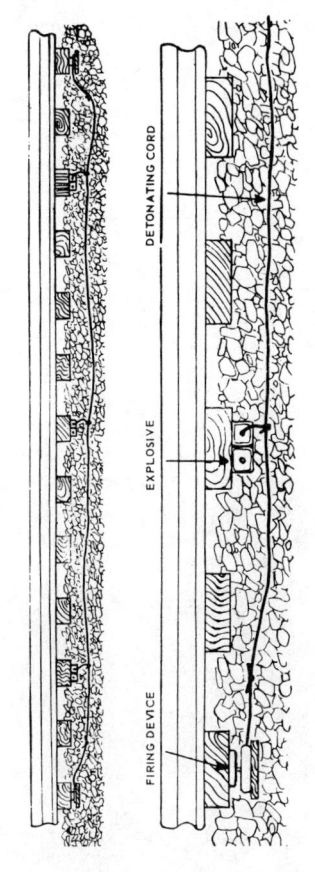

Figure 78. 3-5-2 Technique for train derailment.

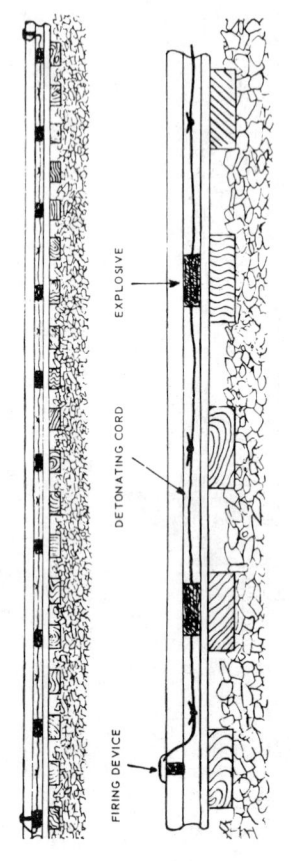

Figure 79. 10-2-1 Technique for train derailment.

(1) *TNT*. TNT is probably the most common explosive. It may be formed in different shapes, but its characteristics are similar to TNT manufactured in the United States since, chemically, they are the same (trinitrotulene).

(2) *Plastic explosives*. Plastic explosives are manufactured by many countries and used for frontline demolition work. Their characteristics and performance are similar to those manufactured in the United States.

(3) *Picric acid*. TNP (Trinitrophenol) is slightly more powerful than TNT with a velocity of about 7,000 meters per second. It is a lemon-yellow, crystalline substance which may be identified by its tendency to dye water or material it may contact. It combines readily with some metals to form picrate (explosive) salts which are extremely sensitive to shock, friction, and heat. For this reason careful attention must be paid to packaging (usually paper or zinc is used) and storage. Otherwise, TNP has the same general characteristics as TNT.

(4) *Guncotton*. The power of guncotton, which is a cellulose of high nitration, is directly influenced by moisture. Dry guncotton generally detonates at a velocity of 7,300 meters per second; when wet,

the velocity of detonation is about 5,500 meters per second. Dry guncotton is extremely sensitive to shock and should be used only for booster pellets and blasting caps.

(5) *Nitroglycerin explosives.* Standard, ammonia, and gelatin (Gelignite) dynamites are common in foreign countries. Granular or free-running dynamite is conventional for borehole loading and replaces black powder in some areas of the world. It usually is less sensitive than other dynamites because of the increase of ammonium nitrate or other compounds necessary to make it pour. Nobel's 808 is similar to blasting gelatin, being of a higher density though somewhat less sensitive. It has a hard, rubberlike texture which tends to soften as the temperature is increased. Its color normally varies between green and brown.

145. Foreign Accessories

a. Primers. Many foreign explosives are as insensitive to shock as TNT. Since most foreign blasting caps are only equivalent to the standard, commercial, American, numbers 6 and 8 caps, the insensitive foreign explosives cannot be detonated consistently by using the American caps. A small amount of a more sensitive explosive must be used as the link between the charge and the cap;

Table III. World's Principal Explosives.

U. S.	British	French	German	Italian	Japanese	Russian
TNT	TNT Trotyl*	Tolite	Pull Pulver Spreng Munition 02.	Tritolo Tritolo*	Chakatsuyaku	TOL Trytyl*
Cyclonite C3* C4*	Plastic explosive or PE-2A.*		Cyclonite* Hexogen C6* Plastite* Nipolit*	Hexagene* T-4*	Koehlenbakuyaku Cyclonite* O-Shizuuyaku*	Hexogen Kaumikite*
Tetryl Tetrytol*	Composition explosive or C.E.				Meisayaku	TETP
PETN Pentolite* Primacord* (Detonating cord)	PETN Pentolite* Cordtex* (Detonating cord)		Knallzundschnur		Shoe-i-yaku**	Ten DSH* 1943**
Ammonium nitrate Amatol	Ammonal* Monobel* (Australian).	Nitrate d'ammonium.	Ammon Saltpeter	Nitrato d'ammonio PNF* Schniderite* Totnal amonal*	Ammon Yaku Shonayaku* Shoan* Gokuyaku*	Gromoboy Ammonite* Dinomonk** Maisite*
Nitroglycerin Dynamite* Blasting gelatin*	Dynamite* Blasting gelatin Geligniter* Nobel's 808*		Dynamite*		Dainamaito*	Graeuline* Dynamon T*
Picric acid (TNP) (no longer used).	Picric acid Lyddite*	Melinite*	Pikrinsaure	Acido picrico Pertite*	Oshokuyaku Shimose* Ochiyaku* Haishoyaku*	Melinyte*
	Gun cotton					Pyroxylin

*Compounded with other explosives.
**Undetermined if this is demolition explosive or detonating cord.

this is called a booster or primer. Foreign demolition charges of the cast kind require the use of a booster and are manufactured with a booster recess.

b. Blasting Caps. Foreign blasting caps are often identical to the American number 6 or 8 caps. They may be of dry guncotton or some other compound pressed into a cardboard, metal, or paper shell. Some of the Russian caps are made of cardboard and paper and may be of slightly different lengths and diameter.

c. Burning Fuse. It is important to recognize instantaneous fuse manufactured by some countries for boobytrapping and incendiary purposes. They burn at fast speeds; some burn as fast as 61 meters per second. When ignited it may appear to explode. To minimize accidents, all fuse should be tested before being used with explosives. Activate unidentified fuze with a firing device from a safe distance or with a known time fuse with a 45° splice.

146. Handling Foreign Explosives

a. General. The characteristics of an unknown explosive must never be taken for granted, and should be subjected to the expedient test methods outlined below.

b. Procedure for Handling Unknown Explosives. Unknown explosives should be tested as follows:

(1) Examine the packaged unit (case, block, cartridge) for exuded liquids. If there

is reason to believe that an oozing explosive is dynamite (i.e., contains considerable nitroglycerin) it should be destroyed.

(2) Subject 0.5 kilogram of the explosive to rifle fire. If it fails to detonate after five or more hits, it may be considered insensitive to shock and friction. Dynamite containing nitroglycerin should detonate on the strike of a bullet.

(3) Place approximately 28 grams of the explosive on paper or some other combustible material and ignite it. This permits the tester to withdraw to a safe distance before the flame reaches the explosive. Take note of the following burning characteristics: color of flame, rate of burning, whether or not the explosive melts, amount and color of smoke, etc. These may be similar to the burning qualities of known explosives and an indication of the content of the unknown explosive compound.

(4) Attempt to detonate a unit of the unknown explosive with a blasting cap. If this fails, increase the number of blasting caps by one for each successive attempt until detonation occurs.

Section II. COUNTERINSURGENCY

147. Counterinsurgency Operations

In support of counterinsurgency operations,

the detachment commander and the combat engineer specialist will place primary importance on those actions designed to win the willing and active cooperation, assistance, and support of the people. In remote areas, where Special Forces detachments will normally operate, there may be a lack of sophisticated structures of any kind. The construction of buildings may well be the assigned mission of the detachment, as opposed to combat operations. Extensive area studies conducted before commitment will reveal additional information on which to prepare plans and details of operations. In preparing for commitment, the engineer specialist will conduct extensive training and development in the field of expedient engineering that may include—

   (1) Road expedients.
   (2) Expedient crossings and bridges.
   (3) Landclearing for farming.
   (4) Construction of lifting devices.
   (5) Construction of simple sanitation projects.
   (6) Use of tools and materials for simple engineering.
   (7) Training and advising indigenous construction and combat engineering units in general construction tasks and in the preparation of defensive fortifications for security of the local villages.

148. Expedient Engineering

Programs undertaken by Special Forces detachments supporting counterinsurgency operations are called civic action or environmental improvement programs. Special Forces detachments conducting military civic actions find that they are the contact, or go-between, for the local administration and the national government. In undertaking these programs and in assisting the local administration to satisfy the aspirations of the people, the Special Forces advisor helps create the image of a responsive and capable government. When this is accomplished, the opening for subversion diminishes.

149. Civic Actions

a. In assessing the capabilities of the units and minority groups advised, the Special Forces commander will propose military civic action projects in accordance with the overall counterinsurgency plan and within the capability of the indigenous units. The Special Forces detachment commander must insure that the objectives of proposed environmental improvement programs will—

(1) Contribute to the betterment of the lives of the local populace.
(2) Gain the support, loyalty, and respect of the people for the government and contribute, in some measure, to national development.

b. The Special Forces detachment undertaking civic action programs must evaluate each program from the standpoint of resources required to complete each task. Harvesting and road improvements, for example, may be undertaken by paramilitary units possessing little more than a labor pool and manpower. The detachment commander and his engineer specialist encourage their counterparts and local population to use local material and equipment as much as possible before requesting assistance from other U.S. support facilities. Where it is required, indigenous engineer troops may be used in tasks requiring a certain degree of skill; but, maximum use of trained personnel should be made from local units. Those tasks requiring pure labor should be relegated to the local villages on a self-help basis. These actions will provide the Special Forces detachment with immediate work on the project and still afford a degree of training to local engineer units to increase their skill levels.

c. In all environmental improvement programs undertaken, Special Forces personnel must insure that the local, indigenous soldier understands that his actions are accomplishing the following objectives:

(1) The soldier is learning his responsibility toward his community.

(2) On interchange of skills between soldier and civilian, there is an exchange of ideas and understanding that enhances national unity.

(3) A soldier learns skills which will be useful in his home village.

(4) Soldiers possessing special skills have the opportunity to increase these skills and

prepare for future employment with local governments as well as with a higher administration.

d. For techniques in the performance of military civic action programs and functions see pages 12 through 19, FM 31–73.

150. Construction Programs

Special Forces detachment personnel may find it necessary to employ the technical skills and capabilities of engineer units of the host country forces for projects supporting environmental improvement programs; however, the Special Forces detachment must adhere to fundamentals and avoid the more advanced techniques and procedures, particularly those that are not compatible with limitations of terrain, road nets, size of host forces, and mobility. Special Forces personnel will try to improvise when standard equipment is not available. The assessment and evaluation of units' and local villagers' capability and availability will dictate those projects to be undertaken. They may include—

a. General Construction Tasks. This may include rough carpentry; construction of drainage facilities with logs and stakes; construction of adobe buildings; rigging, and lashing techniques; and construction of small, water supply reservoirs.

b. Military Engineer Tasks. Here the emphasis will be on field fortifications and protection from direct weapons fire, rather than blasts from heavy artillery and large explosives. Considerations should be given to trench-type fortifications around fixed installations. Additionally, the Special Forces may assist in the preparation and use of—

   (1) *Obstacles.* Preferably anti-personnel obstacles as opposed to vehicular; installations of minefields and barbed-wire; construction of nuisance items such as heavy brush and impaling devices; construction of watch towers; and using natural obstacles to impede vehicular movement.

   (2) *Boobytraps.* Improvised traps for warning devices (FM 5-31); using selected items of clothing and equipment that would naturally appeal to an enemy; and anti-personnel mines employed in normal defensive positions.

   (3) Demolitions used to improve mobility of tracked vehicles by reducing steep banks, destruction of tunnels, and underground hiding places.

c. Specific Construction Projects.

   (1) Construction of bridges and ferries from natural materials.

   (2) Routes of communications which may include construction and improvement of roads, ditching, drainage, and temporary construction of air landing facilities.

(3) Land clearing for agriculture projects. For detailed information on construction programs that may be employed, see FM 31-73.

151. Resources Control

Through extensive training and constant development of destructive techniques, the Special Forces detachment personnel learn the various materials and their many uses in making destructive devices. Through extensive studies of their operational areas, they determine the availability of these materials to the local populations as well as the insurgent force. The Special Forces detachment commander is able to advise his counterparts on resources control measures to deny the insurgent access to such materials. The detachment commander must exploit all available means to help the local law enforcement agencies prevent essential resources from falling into the hands of the insurgent. The police and paramilitary forces in operational areas must be properly oriented and indoctrinated for this task.

a. In establishing requirements for resources control, priorities must be assigned to specific items to be denied the insurgent. Restrictions on certain items may be injurious to the attitude of the population, such as the control of fertilizer in a primarily agrarian area. Two methods may be employed in controlling materials—

(1) *Price regulation.*
(2) *Rationing.*

b. Additional controls must be employed for materials that can be used as expedients in manufacturing improvised explosives. Adequate control of these items will depend upon properly trained, security personnel positioned at the production and distribution facilities for these sensitive items.

    (1) *Physical security.* Physical security could include check points for searching personnel and vehicular traffic entering and leaving installations; detection devices for certain items that react to electronic devices; clothing change points requiring personnel to shower and change clothes on entering or leaving installations.

    (2) *Personnel security.* Personal security is more difficult; however, Special Forces personnel, working in close conjunction with local police and security elements, may instigate a personnel security investigation to insure that personnel selected for work are reasonably clear of implications with known insurgent members. Additional procedures may be—

  (*a*) Planting of informers.
  (*b*) Offers of rewards for information.
  (*c*) Planting of erroneous information concerning activities.
  (*d*) Surveillance of after-duty-hour activities.
  (*e*) Curfews.

  c. The use of resources control measures is sensitive and must be carried out with utmost discretion. Infringement upon the rights of the local population, through violence or needless oppression, will lose the population to the insurgent. Local law enforcement agencies should be closely supervised at all times during the operation.

Section III. METRIC CALCULATIONS

152. General

The following metric formulas may be used for demolition projects when working with personnel familiar with the metric system. Use of metric formulas and construction and placement of charges are the same as for U.S. Corps of Engineer formulas and charges. Since the formula results give kilograms of TNT, the relative effectiveness of other explosives must be considered. For demolition formulas see FM 5–25, or Demolition Card (GTA 5–10–9).

  a. Structural Steel.
  Formula: $Kg = \dfrac{A}{38}$

  Kg = Kilograms of TNT required.
  A = Cross-sectional area in square centimeters.

  b. Timber.
   (1) *External charge.*
  Formula: $Kg = \dfrac{D^2}{550}$

  Kg = Kilograms of TNT required.
  D = Diameter of target in centimeters.

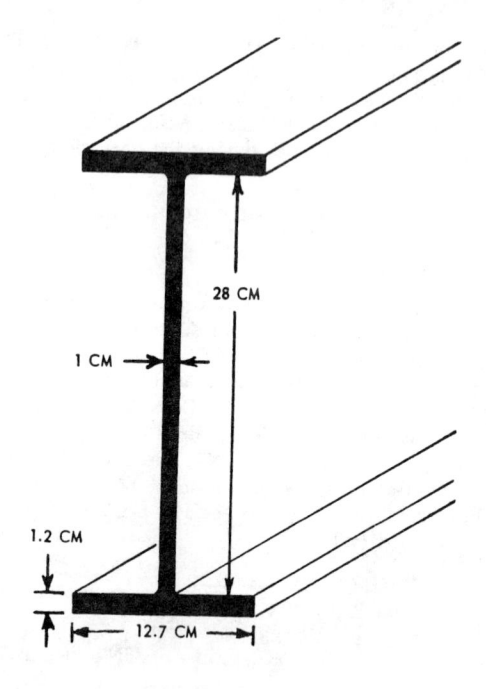

Figure 80. Calculation for cutting steel I-beam.

(2) *Internal charge.*

Formula: $Kg = \dfrac{D^2}{3,500}$

$Kg =$ Kilograms of TNT required.
$D =$ Diameter of target in centimeters.

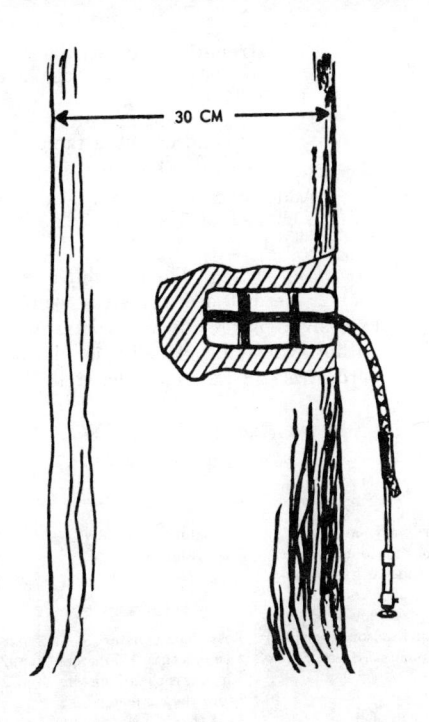

Figure 81. Internal charge to cut timber.

c. *Breaching.*

Formula: $Kg = 16\ R^3KC$
 $Kg =$ Kilograms of TNT required.
 $R =$ Breaching radius in meters.
 $K =$ The material factor based on

strength and hardness of material to be demolished (table IV).

C = The tamping factor based on type and extent of tamping to be used (fig. 82).

Add 10 percent to a calculated charge of less than 22.5 kilograms.

(1) *Breaching radius.* The breaching radius (R) is the distance in meters which an explosive charge must penetrate and within which all material is displaced or destroyed. For example, if it is de-

Table IV. *Material Values of K Factor.*

Material	R	K
Ordinary earth	All values	0.05
Poor masonry, shale and hardpan good timber and earth construction.	All values	.23
Good masonry, ordinary concrete, rock.	Less than 1 meter	.35
	1 to less than 1.5 meters	.28
	1.5 to less than 2 meters	.25
	More than 2 meters	.23
Thick concrete, first-class masonry.	Less than 1 meter	.45
	1 to less than 1.5 meters	.38
	1.5 to less than 2 meters	.33
	More than 2 meters	.28
Reinforced concrete (will not cut reinforcing steel).	Less than 1 meter	.70
	1 to less than 1.5 meters	.55
	1.5 to less than 2 meters	.50
	More than 2 meters	.43

sired to break a 2-meter concrete wall by placing a charge on one side, then the value of R, in the formula $Kg = 16 R^3 KC$, is 2.

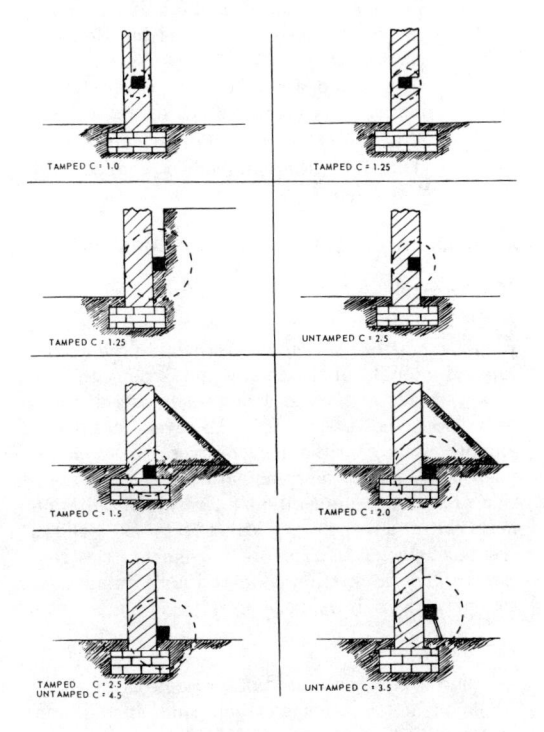

Figure 82. Value of C (tamping factor).

(2) *Material factor* (table IV). The values of material (K) for various types of construction are given in the following tables:

(3) *Tamping factor*. The value of the tamping factor depends on the location and the tamping of the charge. No charge is fully tamped unless it is covered to a depth equal to the breaching radius.

(4) *Number of charges*. For calculations to determine the number of charges, see FM 5-25.

Section IV. ATOMIC DEMOLITION MUNITION

153. General

ADM is employed in conformance with tactical requirements of the assigned mission to reduce the tactical mobility of the enemy and to deny the use of key facilities such as bridges, industrial facilities, and power plants; however casualties among civilian personnel, destruction of manmade and natural terrain features, and the creation of areas of high intensity, residual radiation may cause adverse political effects as well as create obstacles to friendly movement. Destruction and contamination is held to a minimum consistent with military necessity.

154. Procedures

a. For command and staff procedures in ADM employment to include troop and installation safety requirements, see FM 5-26.

b. See FM 31-21A, Special Forces Operations for information on personnel to employ ADM, target selection and coordination, target analysis, preparation for ADM mission, logistical procedures, employment, and support of conventional forces employment of ADM.

c. Firing option, emplacement consideration, and nuclear effect data are contained in FM 5-26A Employment of ADM (U).

d. Operational techniques—see TM 9-1100-205-12.

e. For a suggested SOP for employment of ADM, see appendix IX.

CHAPTER 12

MEDICAL ASPECTS OF SPECIAL FORCES OPERATIONS

Section I. GENERAL

155. General Medical Requirements

In efforts to gain the support of local populations, medical care has proven to be a most effective instrument. The offer of medical assistance may be used effectively to achieve entry into hostile areas and to assist in gaining the support of those indigenous populations with indifferent (or undecided) loyalties; however, a basic distinction exists between unconventional warfare and counterinsurgency operations with respect to the goals one hopes to achieve in this manner. The goal of medical operations in the unconventional warfare situation is to secure the support of local populations for U. S. forces operating within the GWOA. In counterinsurgency operations, the goal is to attract the loyalties of the villager to the central government. It is well to point out that in certain areas of the world the local population may not be receptive to the Western medical practices and concepts. Before a medical program can be initiated in such an area, it will be necessary to persuade the populace to accept the program.

156. Organization for Medical Support

a. The organization of medical elements in unconventional warfare or counterinsurgency operations is tailored to fit the particular situation. The basic medical organization is organic to the Special Forces group; however, it may be augmented by personnel from medical augmentation detachments, depending upon the skills required. In any case, this basic medical organization will be expanded, as appropriate, through use of trained, indigenous, medical personnel and through implementation of medical training programs for indigenous civilian, military, and paramilitary personnel.

b. Skills organic to the Special Forces medical organization provide for the following capabilities:

   (1) The provision of organizational medical care, medical supply, and dental service to the group, or elements thereof, and to indigenous military and civilian personnel in consonance with command policy.

   (2) The planning, supervision, and conduct of programs for the instruction of U.S. forces and indigenous military and civilian personnel in—

   (*a*) The care of casualties from disease and injury.

   (*b*) Personal, organizational, and community measures for the preservation of health.

- (c) The selection and preservation of foods.
- (d) The care and handling of pack animals and on field expedients in rigging packs and litters for animal transport.

(3) The provision of preventive medicine functions, medical technical intelligence, and veterinary activities to include—
- (a) Epidemiologic investigation of conditions affecting the health of U.S. forces, indigenous military and civilian personnel, and animals.
- (b) Field surveys and inspection of significant environmental factors affecting the transmission of disease.
- (c) The planning and application of measures to control diseases and disease reservoirs in U.S. forces and indigenous military and civilian personnel.

157. Preventive Medicine Techniques

Preventive medicine techniques applicable to Special Forces operations may be divided into four phases—

a. Procedures in garrison and during field training, before deployment in unconventional warfare or counterinsurgency operations.

b. Procedures in unconventional warfare operations.

c. Procedures in counterinsurgency operations.

d. Procedures for collecting medical intelligence from areas of unconventional warfare or counterinsurgency operations.

158. Preventive Medicine Procedures Prior to Deployment

a. During periods of garrison duty and field training, preventive medicine activities in Special Forces units are directed toward maintaining the highest standards of personal hygiene and cleanliness in troop areas and facilities, unit and individual training in preventive medicine techniques, and the preparation of area medical studies. Technical advice and supervision are provided in connection with—

(1) Food and its preparation.
(2) Water supply.
(3) Troop housing.
(4) Bathing and latrine facilities.
(5) Waste and garbage disposal.
(6) Insect and rodent control.
(7) Sanitation in campsites.

b. Periodic reports are prepared to keep the commander informed of the status of the health of his command of the conditions which may adversely affect health. Corrective action is recommended for unsatisfactory conditions.

c. During the period immediately preceding deployment, preventive medicine activities are directed primarily toward—

(1) Briefiings on the general medical situa-

tion in areas of planned deployment, to include information on the endemic diseases and on individual and small-unit measures which may be implemented for their prevention.

(2) Immunizations, to provide a high degree of immunity to certain common disease conditions.

(3) Predeployment medical examinations, to identify and eliminate those personnel with medical conditions for which treatment in remote situations would be difficult or impossible.

Section II. MEDICAL REQUIREMENTS FOR GWOA

159. General

a. Medical requirements within the GWOA will differ in two respects from those posed by conventional operations.

(1) Battle casualties are normally lower in guerrilla units than in conventional units.

(2) The incidence of disease is often higher in guerrilla forces than in conventional units of similar size.

b. The medical organization in support of guerrilla forces will ordinarily feature both organized medical groups and auxiliary medical facilities. The former are usually located in guerrilla base areas and staffed by guerrilla medical detachments. Auxiliary facilities are in locations in which individual patients (or a small number of patients) may be held in a convalescent status, or may be sustained until a time when it is safe to evacuate them to more advanced treatment facilities in "safe" areas.

160. Evacuation and Hospitalization in the GWOA

a. Evacuation.

(1) Since evacuation within and from the GWOA is normally difficult, unit commanders must rely on their own resources and on support from auxiliaries and the underground in planning the evacuation of casualties. Great reliance must be placed on self-aid. Maximum use must be made of specially trained, enlisted, medical personnel who may give treatment which obviates the requirement for evacuation. Local pack animals and other civilian ground and water transportation should be used to the maximum possible extent, as well as litter bearers recruited from among indigenous personnel. Every effort is made to evacuate wounded personnel from the scene of action. The condition of the wounded and the tactical situation may preclude the transportation of casualties along with the unit to the guerrilla base. In this event, the wounded may be hidden in safe sites or

well-concealed locations. The auxiliary, who can care for the wounded until their return to active duty is possible, is notified. All planned operations should include an SOP for emergency treatment and evacuation. When tactically possible, use should be made of scheduled and on-call air evacuation from the area.

(2) The removal of dead from the scene of action is most important for security reasons. Identification of the dead by the enemy may jeopardize their families and their units. The bodies of those killed in action are removed and cached until recovery is possible. Bodies are then disposed of by means consistent with the customs and religions of the local population.

(3) As the overall tactical situation begins to favor the sponsor, evacuation of sick and wounded to friendly areas may become feasible. This lightens the burden upon the meager facilities available to the area command and provides a higher standard of medical care for the patient.

b. Hospitalization. The care and treatment of sick and wounded will generally be accomplished by guerrilla medical personnel within the area, until evacuation of selected personnel can be accomplished to friendly areas outside the GWOA **through the auxiliary and the underground. Small, isolated, and well-hidden treatment and** holding facilities may be established. When necessary, the auxiliary and the underground may assist in the infiltration of medical personnel and equipment to accomplish life-saving procedures. In some instances, the auxiliary may arrange hospitalization in widely-scattered, private homes where periodic visits can be made by medical personnel. As the GWOA expands, the services of professional medical personnel and the facilities available in villages and towns within the GWOA may be available during certain hours, if not for complete and continued hospitalization.

161. The Build-Up Phase of Unconventional Warfare Operations

a. During the build-up phase of unconventional warfare operations, an initial assessment is conducted to determine the state of sanitation and health within the GWOA. Such an assessment will include consideration of the diseases endemic to, as well as potential epidemic diseases within, the area of operations.

b. The guerrilla force will be comprised of both foreign and indigenous personnel; therefore, differences in immunity to the endemic diseases will exist between these two groups. Certain preventive measures may be applicable to one group and not to the other. This factor must be considered in the implementation of immunizations and use of chemoprophylactic agents, for instance, the use of chloroquine-primaquine in the prevention of malaria. (While it may be necessary to administer routine prophylaxis to those recently introduced to the area, it may not be desirable to treat indigenous guerrilla forces with chemoprophylactic drugs, since such treatment may alter naturally-acquired immunity to this infection.) Further, the indigenous members of the force may exhibit diseases uncommon to U.S. forces, such as serious nutritional deficiencies and cases of active tuberculosis.

c. It is during the build-up phase that the guerrilla force is thoroughly indoctrinated in preventive medicine procedures which must be strictly adhered to during active operations. This indoctrination will be directed toward basic sanitation, personal hygiene, and individual protective measures. Specific attention will be given to the use of chemoprophylactic agents, immunizations, food and water sanitation, individual methods for protection against bites from arthropods, and the sanitary disposal of human wastes. Recommendations are made to individual commanders with respect to the measures to be enforced in all localities of the GWOA. The scope of the preventive medicine effort must be adequate to encompass the probable expansion of the guerrilla force.

d. The preventive medicine section organic to the Special Forces group will conduct a principal preventive medicine area study and assessment to be used as a guide to planning and implementing preventive medicine techniques appropriate to all phases of unconventional warfare operations. This area study and assessment will consider the preventive medicine requirements for the indigenous civilian population (i.e., the dependents of the guerrillas), as well as the guerrilla force itself and should include the material suggested in appendix V.

162. The Employment Phase of Unconventional Warfare Operations

During the period of employment of the guerrilla force in active operations against the enemy, preventive medicine activities must be directed toward the prevention of disease among individual members of small, highly mobile, operational units which will be deployed throughout the GWOA. Preventive medicine techniques will be determined by local and immediate requirements. Preparations must be made to meet unexpected problems such as widespread epidemics. Preventive medicine programs will be limited, due to the nature of guerrilla operations, to individual and small-unit measures for the prevention of disease; however, the programs must include the civilian populations of the villages from which the guerrilla forces originate and from which local support may be expected. The support of the civilian population may be enhanced by offers of medical assistance. Members of the families of guerrilla forces must be actively encouraged to adopt the same standards of personal hygiene and sanitation as those enforced among the guerrilla units.

163. The Demobilization Phase of Unconventional Warfare Operations

a. The demobilization phase occurs when juncture between friendly, conventional forces and the area command is completed, and the ability of the guerrilla forces to support the military operations gradually diminishes. This phase is characterized by social disorganization created by relocation and resettlement of large numbers of displaced persons, refugees, and evacuees. Disorganization of civil government, the disruption of public utilities, and mass migrations all contribute to the potential for widespread epidemics.

b. The role of the preventive section organic to the Special Forces group now changes from one of support for the guerrilla unit and the families of the guerrillas, to one of assisting civil government and military civil affairs units to implement public health measures within the areas of guerrilla operations. The principles set forth in the paragraphs on counterinsurgency operations are, in general, applicable to this phase.

164. Medical Supply in the GWOA

a. In all probability, medical supplies in the GWOA will be available in limited quantity. Preplanning to provide the minimum essential medical supplies and equipment for current, planned, and contingency operations is mandatory. During the area assessment, and as the development of the guerrilla medical organization progresses, medical supplies are requested from the SFOB by the operational detachments based upon their operational requirements. Excess medical stock will be maintained at the SFOB to expedite prompt resupply or delivery of medical items to operational detachments with the greatest requirements. Medical supplies may include dental equipment, blankets, drugs, bandages, and ambulatory aids. As the medical facility expands there may be a requirement for special items of surgical equipment. Medical items are ordered through use of the Catalog Supply System (app VII).

b. There should be minimum dependence on the local economy for the provision of medical supplies and equipment, because such items will certainly be scarce and in great demand by the local population. On the other hand, captured medical supplies are of value in augmenting guerrilla stocks. Such materials should be returned to the SFOB as expeditiously as possible for redistribution according to overall operational requirements.

c. It may be necessary to cache excess medical supplies in order to maintain mobility and deny access to the enemy. Precautions must be taken to prevent spoilage.

d. Medical supplies are strictly controlled by the area command, since such articles are potential blackmarket items.

Section III. MEDICAL REQUIREMENTS FOR COUNTERINSURGENCY OPERATIONS

165. General

Requirements in counterinsurgency operations differ from those in the GWOA, in that medical activities are conducted openly, using existing medical organization and facilities. The medical organization should provide for training and operational assistance to indigenous military and paramilitary forces of the host country, with particular emphasis on the development of civic action programs. Civilian personnel, selected from the community, are trained, in cooperation with existing civilian health agencies and U.S. AID missions, to improve health and sanitary conditions in local villages (so-called village health workers). These personnel are trained in basic first aid and health and sanitation, either at centrally located medical training facilities or in the village. These indigenous personnel will carry out self-help programs in sanitation within the village under the supervision, and with the advice, of medical technical personnel organic to the Special Forces group or special action force.

166. Preventive Medicine in Counterinsurgency Operations

a. In contrast to unconventional warfare operations, in which primary emphasis is on those measures which will improve and maintain the health of the guerrilla unit, the effort in counterinsurgency operations is directed toward improvements in health and sanitation among indigenous civilian populations. The general steps to be taken in implementing such programs are—

(1) Establish liaison with existing health authorities.

 (2) Accomplish an initial area assessment.
 (3) Attempt to secure the support of the village leaders.
 (4) Implement a training program for village health workers.
 (5) Implement health and sanitation measures based on priorities and the desires of the villagers.

b. The success of counterinsurgency operations at a village level requires tangible evidence that the central government is responsible for efforts to improve the lot of the villager. Early liaison with appropriate local representatives of existing health agencies is, therefore essential in order to achieve support, approval, and participation in plans for health programs to be implemented in the village.

c. The success of health programs will depend largely upon one's ability to motivate the villager to undertake changes in habits which have been practiced for generations. To accomplish this, it is necessary to have a knowledge of the social structure of the village (Who are the official and unofficial community leaders?) and the local beliefs, customs, taboos, and mores. (In many primitive societies the occurrence of disease is associated with visitations by evil spirits.) The collection of information of this nature is part of the process of area assessment. Once some insight has been acquired into these matters, it is usually possible to lay out an intelligent plan by which to attack basic health problems in the village.

d. Efforts are then directed toward motivating and training local villagers to accomplish these objectives. Local support is usually best achieved through the village leaders (the village council or similar governing body), to include the unofficial leaders (opinion formers) who, although not acting in an official capacity, nevertheless exert great influence within the community.

e. Training programs in basic health subjects must be initiated for individuals who can successfully use their knowledge to help the people help themselves toward better health. It is usually wise to permit the village council to select those who will undertake such training. The pretige associated by the villagers with this activity will usually result in the selection of individuals who are already in a position of influence within the community, thereby giving additional emphasis to the program. The subjects recommended for such training programs are—
  (1) Germs and parasites as causes of disease.
  (2) Food and water sanitation.
  (3) Personal hygiene.
  (4) Village sanitation, latrine, and bath facilities.
  (5) Pre- and post-natal care.
  (6) Nutrition and health.
  (7) Arthropod and rodent-borne disease control.

f. Although initial emphasis is placed on enlisting the support of selected leaders and training village health workers, it is also necessary to obtain active participation by the villagers in order to accomplish the goals of improving and maintaining village health. Some suggested projects for general village participation might include one or more of the following:

(1) General village improvement teams.
(2) Waste disposal inspection teams.
(3) Food and water inspection teams.
(4) Rodent and vector control teams.

g. Priorities for programs are based upon the initial area assessment as well as upon the desires ("felt needs") of the community. In initiating programs, give consideration to these basic rules—

(1) Know the community and its leaders.
(2) Do not unnecessarily interfere with the people's customs.
(3) Get the people to help themselves.
(4) Keep programs simple and practical.
(5) Build and maintain momentum.
(6) Build trust.
(7) Make it fun and convenient.
(8) Plan for permanence.

Section IV. VETERINARY MEDICAL TECHNIQUES

167. General

Veterinary medical techniques applicable to Special Forces and special action force operations may be divided into four phases—

a. Procedures in garrison and during field training, before deployment in unconventional warfare or counterinsurgency operations.

b. Procedures in unconventional warfare operations.

c. Procedures in counterinsurgency operations.

d. Procedures for the collection of veterinary medical intelligence from areas of unconventional or counterinsurgency operations.

168. Veterinary Procedures Prior to Deployment

a. During periods of garrison duty and field training, veterinary activities are directed toward the preparation of area medical studies and individual and unit training on the subjects of—

 (1) Wholesomeness and sanitation of subsistence.
 (2) Care and management of pack animals.
 (3) Food inspection procedures.
 (4) Zöonotic diseases.
 (5) Techniques for using animals for pack and transportation.
 (6) Survival techniques.

b. Immediately before deployment, veterinary activities are directed toward—

 (1) Briefings on the veterinary medical situation in areas of planned deployment, to include endemic and potential epidemic zöonoses, and on individual and

small-unit measures which may be implemented for their prevention.

(2) Briefings on diseases of animals in areas of planned deployment that may directly or indirectly influence the outcome of unit deployment.

169. The Build-Up Phase of Unconventional Warfare Operations

a. During the build-up phase of unconventional operations, veterinary activities will include area studies which are designed to determine the veterinary requirements for support of combined (U.S. and indigenous) guerrilla forces within the GWOA. Area of emphasis should include—

(1) Food and rations and the nutritional requirements of indigenous guerrilla personnel.

(2) Animal diseases transmissible to man.

(3) Availability of animals for transportation and evacuation.

b. Programs based upon this information will usually be implemented as expansion of the guerrilla force occurs and will include—

(1) The establishment of suitable facilities in which to receive, store, and issue rations.

(2) The establishment of standards of acceptability for partisan-supplied foods.

(3) Stockpiling U.S.-supplied rations and the

supervision of preparation of operational rations. (Composition of rations will be based upon the previous assessment of nutritional requirements of guerrilla personnel.)

(4) The establishment of preventive medicine procedures for the control of animal and zöonotic diseases.

(5) The procurement of pack animals.

(6) Training for guerrilla personnel in—
 (a) Survival techniques to be used by individuals and small, operational units.
 (b) The selection and preparation of indigenous foods.
 (c) The care and handling of pack animals.

c. Recommendations will be made to appropriate guerrilla commanders with respect to veterinary preventive measures which must be initiated and enforced to control endemic and potentially epidemic diseases of animals, and the zöonotic diseases.

170. The Employment Phase of Unconventional Warfare Operations

a. During this phase, veterinary activities will consist of continued evaluation of veterinary data and the formulation of plans to improve environmental sanitation, the provision of food supplies, and the control of animal diseases and zöonotic conditions within the GWOA.

b. Emphasis will be on providing assistance and technical information to deployed operational detachments.

171. The Demobilization Phase of Unconventional Warfare Operations

a. During the demobilization phase, veterinary activities are redirected toward providing an adequate food supply for large numbers of displaced persons, refugees, and evacuees. Secondary efforts are directed toward the control of animal diseases of public health significance.

b. The major veterinary programs, formulated at theater level, will be directed at redevelopment of food production and processing and will be implemented under the operational control of the civil government and the civil affairs units having jurisdiction within the country.

Section V. COLLECTION OF INTELLIGENCE AND INFORMATION

172. General

a. Intelligence collecting is an inherent capability of medical personnel. The activities of medical personnel in treating members of the local population and administering to the sick and wounded insurgent provide innumerable opportunities to collect intelligence; for example, information on the effects of environmental improvement program; the efforts of propaganda on the populace; and information on weapons, equipment, medical supply, and morale. This function is over-and-above that of collecting technical medical intelligence. The deployment of small units to such areas provides a unique opportunity to delineate the military disease problems of the area by using deployed personnel as "sentinels."

b. An appropriate battery of screening examinations can usually be devised for a given area which is administered before deployment to obtain baseline data. The same examinations accomplished on the return from a mission serve the dual functions of detecting those individuals who have acquired disease during the mission which requires treatment, and delineation of the major disease problems of an area by systematically tabulating the results of examinations and mapping the results according to the probable area of acquisition of disease. This effort requires the support of the advanced, medical laboratory facilities in rear areas.

c. Epidemiologic surveillance, conducted in this manner, serves two functions—

(1) It provides the basis for recommending preventive measures to be taken by units to be deployed in these areas in the future.

(2) It brings to attention those disease problems of major military importance which require further investigation within operational areas by teams of

FM 8-50	Bandaging and Splinting.
FM 8-55	Army Medical Service Planning Guide.
FM 19-40	Handling Prisoners of War.
FM 20-32	Land Mine Warfare.
FM 21-5	Military Training.
FM 21-6	Techniques of Military Instruction.
FM 21-10	Military Sanitation.
FM 21-11	First Aid for Soldiers.
FM 21-20	Physical Training.
FM 21-30	Military Symbols.
FM 21-50	Ranger Training & Ranger Operations.
FM 21-60	Visual Signals.
FM 21-76	Survival.
FM 21-77	Evasion and Escape.
(C) FM 21-77A	Evasion and Escape (U).
FM 24-16	Signal Orders, Records, and Reports.
FM 24-18	Field Radio Techniques.
FM 30-5	Combat Intelligence.
FM 30-7	Combat Intelligence-Battle Group, Combat Command, and Smaller Units.
FM 30-9	Military Intelligence Battalion, Field Army.
(C) FM 30-15	Intelligence Interrogations (U)
FM 30-16	Technical Intelligence.
FM 30-28	Armed Forces Censorship.
FM 31-8	Medical Service in Joint Oversea Operations.
FM 31-10	Barriers and Denial Operations.

FM 31-16	Counterguerrilla Operations.
(C) FM 31-20A	Special Forces Operational Techniques (U).
FM 31-21	Special Forces Operations.
(S) FM 31-21A	Special Forces Operations (U).
FM 31-22	U.S. Army Counterinsurgency Force.
(C) FM 31-40	Tactical Cover and Deception (U).
FM 31-72	Mountain Operations.
FM 31-73	Advisor Handbook for Counterinsurgency.
FM 33-1	Psychological Operations.
FM 41-10	Civil Affairs Operations.
FM 57-35	Airmobile Operations.
FM 57-38	Pathfinder Operations.
(S) FM 100-1	Doctrinal Guidance (U).
FM 101-31-1	Staff Officer Field Manual; Nuclear Weapons Employment.
(S) FM 101-31-2	Staff Officer Field Manual; Nuclear Weapons Employment (U).
FM 101-5	Staff Officers' Field Manual; Staff Organization and Procedure.
FM 110-101	Intelligence Joint Landing Force Manual.
FM 110-115	Amphibious Reconnaissance; Joint Landing Force Manual.
TM 5-280	Foreign Mine Warfare Equipment.
TM 5-632	Insect and Rodent Control.

TM 8-230	Medical Corpsman and Medical Specialists.
(S) TM 9-1100-205-12	Operator and Maintenance Manual (SADM). (U).
TM 9-1345-200	Identification, Care, Handling and Use of Land Mines.
TM 9-1385-9	Explosive Ordnance Reconnaissance.
TM 9-1385-Series	Explosives Ordnance Disposal.
TM 9-1900	Ammunition; General.
TM 9-1910	Military Explosives.
TM 9-1375-200	Demolition Materials.
TM 10-500-95	Air Drop of Supplies and Equipment: Simultaneous Airdrop of Quantities of A-21 and A-7A Containers.
TM 10-501-1	Army Parachutes Packing Troop-Back Personnel Parachutes (T-10 & Maneuverable).
TM 11-5820-502-20P	Radio Set, AN/GRC-109.
TM 11-296	Radio Set, AN/PRC-6.
TM 11-4065	Radio Sets, AN/PRC-8, AN/PRC-9, and AN/PRC-10.
TM 11-666	Antennas and Radio Propagation.
TM 11-486-6	Electrical Communication Systems Engineering: Radio.
TM 11-5122	Direct Current Generator, G-43/G.

(CM)TM 32-220	Basic Cryptography. (U)
TM 57-210	Air Movements of Troops and Equipment.
TM 57-220	Technical Training of Parachutists.
DA Pam 21-81	Individual Training in Collecting and Reporting Military Information.
DA Pam 108-1	Index of Army Motion Pictures, Film Strips, Slides, Tapes, and Phono-Recordings.
DA Pam 310-Series	Military Publications Indexes (as applicable).
ACP 121	Communication Instructions, General.
ACP 122	Communication Instructions, Security.
ACP 124	Communication Instructions, Radio Telegraph.
ACP 131	Communication Instructions, Operating Signals.

APPENDIX II
FIELD EXPEDIENT PRINTING METHODS

1. Instructions for Making and Using the Silk Screen

a. Tools for the Job. There are six items of equipment which are necessary for printing in the field. The use of these six tools will make it possible to have printed matter available for use at any time and anywhere. The field expedient printer can carry these items along whenever he expects to do printing in the field; however, it is important to remember that a good workable substitute can be found for all of these items in the forests, swamps, and deserts of the world. The field expedient printer can often do his job through the use of substitute items. The six essential tools for printing in the field are—

(1) A silk screen.
(2) A stencil.
(3) Ink.
(4) A stylus.
(5) Paper.
(6) A squeegee, or ink roller.
   (*a*) The silk screen (fig. 83) consists of a frame over which is stretched a piece of cloth. This frame is attached to a base to provide a flat working space.

The cover is not necessary for printing but simply makes the silk screen easy to carry from one place to another.

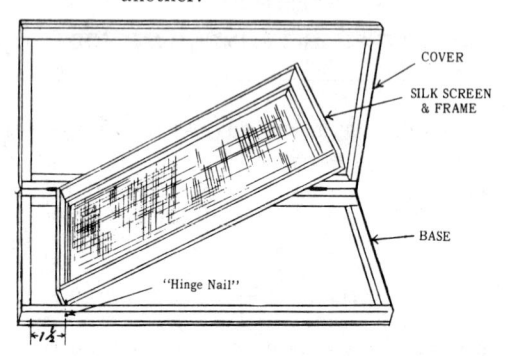

Figure 83. Silk screen with carrying case.

(b) The stencil is a device which allows the ink to pass through the screen and onto the paper where it is needed and blocks out the ink where it is not needed.

(c) The ink used in silk screen printing should be thick and have an oil base; many kinds of ink can be used for printing in the field.

(d) A stylus is a device used to etch the stencil. A pointed piece of wood or metal can be used for this purpose.

(e) Paper or a good substitute is an essential item for printing in the field.

Many good substitutes for paper have been found, but it is best to have a good supply of paper whenever possible. Often paper which has been used can be reused by the printer for a new mission.

(f) The squeegee, or ink roller, is a tool used to spread the ink evenly and to force the ink through the stencil and onto the paper.

b. *Making a Silk Screen.* The field expedient printer can construct a silk screen printing press by following the instructions below. Remember that the silk screen and all of the other items mentioned can be made by using materials found in the field. A good serviceable silk screen can be made by using wooden pegs instead of nails, a rock instead of a hammer, a knife instead of a saw, and bamboo instead of pieces of wood for the frame. Three tools used for making a silk screen are—

(1) A hammer or heavy object for driving tacks and small nails.

(2) A knife for cutting the cloth and canvas hinge.

(3) A saw or hatchet for cutting the wood.

Materials for constructing the frame (fig. 84) are—

(1) 4-pieces of wood 3.18 cm. x 1.91 cm. x 38.74 cm.

(2) 4-pieces of wood 3.18 cm. x 1.91 cm. x 53.34 cm.

(3) 16–2.54 cm. nails.
(4) 2–3.18 cm. nails.

Nails must be very thin so that they will not split the wood. It is best to use "soft" wood in making the frame.

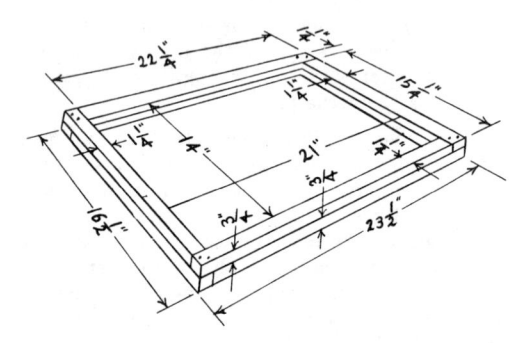

Figure 84. Dimensions for construction of silk screen frame.

When you have made the frame, you are ready to attach the cloth. Many kinds of material can be used to make the screen. Silk cloth is a material which gives the best results; is strong and can be cleaned and used many times. Parachute nylon or a cotton handkerchief will also serve in an emergency and even an undershirt can be used; however, remember that only finely woven cloth will allow a fine line to be printed.

c. Directions for Attaching the Cloth to the Frame (fig. 85).

(1) Cut the piece of cloth so that it is several inches larger than the dimensions of the frame.

(2) Soak the cloth in water so that it will shrink tightly over the frame when it dries.

(3) Place the cloth over the wooden frame and place one tack in each corner as shown in A, fig. 85. Either small .64 cm. tacks or staples can be used. You will need about 90 tacks or staples to attach the cloth securely.

(4) Next, place a row of tacks along one side of the frame as shown in B, fig. 85. Ten evenly spaced tacks or staples will be enough.

(5) Place a row of tacks along the opposite edge of the frame as shown in C, fig. 85. The cloth must be pulled very tightly before driving each of these 10 tacks.

(6) Continue to drive the tacks around the outside of the frame.

(7) Add a second row of tacks around the inside as shown in D, fig. 85. This will give added strength to the screen.

d. Materials for Constructing the Base and Cover (fig. 86).

(1) 4-pieces of wood 2.54 cm. x 2.54 cm. x 43.18 cm.

(2) 4-pieces of wood 2.54 cm. x 2.54 cm. x 71.12 cm.

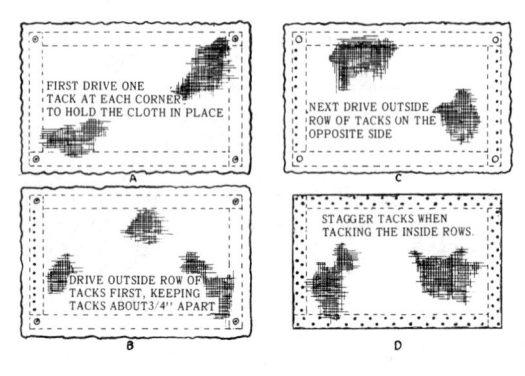

Figure 85. Tacking cloth to underside of frame.

    (3) 2-pieces of cardboard or plywood 48.26 cm. x 71.12 cm.

    (4) 1-piece of canvas or very heavy cloth 5.08 cm. x 71.12 cm.

    (5) 8-3.18 cm. nails.

    (6) 140-.64 cm. tacks.

e. Directions for Making the Base and Cover.

    (1) The four pieces of wood (2 pieces, 43.18 cm. and 2 pieces, 71.12 cm.) are nailed together as shown in figure 86. Two nails are used at each corner.

    (2) The piece of cardboard or plywood is then placed over the wood frame and tacked around the edge with tacks. Space the tacks evenly one inch apart.

f. The Hinge Nails. You are now ready to hinge the silk screen to the base. Place the frame in the

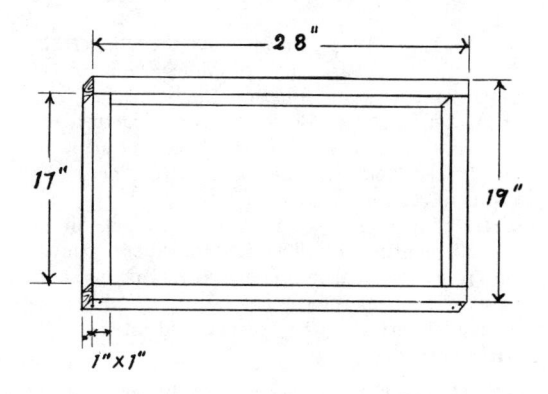

Figure 86. Dimensions for construction of base and cover.

base with the cloth side down. The silk screen is now flat against the cardboard or plywood. The end of the silk screen frame should be 3.81 cm. from the end of the base. This will permit the frame to be raised. The two 3.18 cm. nails are driven through the side of the base from the outside and into the end of the silk screen frame. Figure 83 shows where these hinge nails are placed. These two nails form a hinge which allows the screen to be raised and lowered. The final step in making your silk screen is the hinging of the base to the cover. This is done by using the 5.08 cm. x 66.04 cm. piece of canvas as a hinge. This piece of cloth is tacked along one side of the base and cover. You now have a carrying case for the **silk screen**, making the screen portable.

g. The Ink to be Used. Many different kinds of

ink can be used for printing with the silk screen. Ink with an oil base, such as mimeograph ink, is best. Paint with an oil base is the best substitute, or printer's ink can also be used for field expedient printing. Ink that is used for silk screen printing should be thick; oil base paints are almost the right thickness. A little practice with the silk screen will teach the printer what to look for in a good printing ink. The field expedient printer can practice by using many kinds of ink and paints. In an emergency, berries or any stain producing material can be crushed and an ink substitute produced.

h. How to Use the Stencil and Silk Screen.

(1) The first step is to make sure that you have all six of your tools. They should be clean and in good working order, and you should have enough paper to finish the job.

(2) Place the words, pictures, or symbols on the stencil. If you are using the standard printing stencil, scratch the words onto the stencil with the pointed stylus. If you are using the cut out stencil, remove the parts of the stencil where you want the ink to pass through. Use a knife or sharp object for this purpose.

(3) Lift the silk screen frame up from the base as in figure 83. Place the stencil on the bottom of the screen. Tacks, tape, or glue can be used to hold the stencil in place.

(4) Place a piece of paper on the base under the stencil. This piece of paper will protect the base from ink while you are preparing to print.

(5) Lower the silk screen onto the base. Place enough ink on the silk to cover the screen. Use the squeegee to spread the ink evenly and to force the ink through the opening in the stencil. The squeegee must have a straight edge; another tool which will do the same job is a roller. A roller made of hard rubber is best for spreading the ink on the silk screen. A stiff brush is another tool which can be used for this job.

(6) You are now ready to print. Place the piece of paper to be printed on the base and lower the silk screen on top of the paper. Slide the squeegee firmly over the silk, forcing the ink through the stencil; lift the screen, remove the paper, and allow the paper to dry. If the printing is not dark enough, add more ink to the screen.

(7) When the printing job is finished, remove the stencil and clean the screen and all of the other tools. Also, be sure that the squeegee is very clean.

2. Instructions for Making and Using the Rocker-Type Mimeograph Machine

a. General Instructions. Cover any smooth,

curved surface with a heavy (thick) porous fabric. Saturate fabric with mimeograph ink. Cover ink pad with desired stencil and apply to appropriate paper with a rocker-type movement of the apparatus.

b. Specific Instructions. A frame or base for this aid can be created, on the spot, by using many ordinary items. A wooden block, tin can, glass bottle, can be used as a frame. The machine can be made with crude tools; or, in some cases, the article may be used as it is. A frame may be made from a wooden block, using a chopping axe and a penknife. The surface can be rubbed smooth against a concrete wall or a smooth stone. The block can also be hollowed out to carry ink, styli, and stencil paper for supply purposes. Size can be increased by fastening a piece of sheet metal to the block.

(1) A coat or blanket can yield thick, porous fabric; or felt or burlap can be used. A cover also may be made of many layers of thin fabric. Wrap the fabric around the smooth, curved surface of the printing frame to make an ink pad. The pad can be held in position with tape, string, thumb tacks, or glue.

(2) Saturate the pad with mimeograph ink. This "ink" can be a composite of almost any grease and carbon scraped from a fireplace or grating. Color can be achieved by mixing pigments of color to the grease instead of carbon. Mimeo-

graph ink, commercial grade, is a universal item and is available in any civilized country. Shoe polish, thinned with kerosene or other solvent, is generally available and usable.

(3) Stencils can be made from thin, tough tissue or thin air mail paper by applying a coat of wax (paraffin) to one side. This wax can be rubbed on, then gently warmed to insure uniformity of thickness and penetration of the paper. Only partial penetration is desirable; not saturation.

(4) For a stylus, you may use a ballpoint pen, a slender stick of hard wood, or even a heavy piece of wire with the ends rounded and smoothed enough to etch the wax without tearing the paper. The stylus is used to inscribe the desired message or to sketch on the wax coating of the paper. The paper is then applied to the ink pad with the wax next to the ink. Some of the ink will penetrate through the lines made by the stylus, thus "printing" the blank paper. The undisturbed wax prevents ink's penetrating the paper in unwanted places.

(5) If no mimeograph paper is available, substitute paper chosen for printing should be of quality similar to newsprint; but, almost any paper will suffice.

3. Instructions for Making and Using a Gelatin Printing Device

a. General Instructions. This reproduction method is more commonly known as the hectograph technique, a commercial technique used worldwide. All necessary materials are commercially known by the name "Hectograph" and are available in several variations, from gelatin plates to prepared plates which are fiberbacked wraparound models for machine use (Ditto). The Ditto machines are similar in appearance to mimeograph machines. Emergency or field conditions will probably dictate the use of the simple gelatin plate described below.

b. Specific Instructions. Gelatin, the base for this technique, can be purchased as a Hectograph product, made from gelatin powder produced by food concerns (such as Knox), or made by boiling the bones and skin of animals. (Pulverizing the bone will speed the boiling down process.) Enough gelatin powder should be added to make a semi-solid plate. The warm, liquid gelatin is poured into a shallow, wide container or on a tabletop where it is allowed to cool and set. When properly prepared, it becomes a glass-smooth plate which feels like sponge rubber to the touch. This will be soft enough to absorb the ink but firm enough not to bleed the ink on the master copy. The addition of a little animal glue will toughen the plate and a little glycerine will keep it from drying out too quickly. The effects of these additions are in direct proportion to the quantity used; both are desirable, but not absolutely necessary. Both should be added and well mixed during the liquid stage of the gelatin.

   (1) The master copy is made on a good grade of smooth, tough, hard finish paper. The material to be reproduced is typed or written using *Hectograph* or *Ditto* carbon paper, ribbon, ink, or pencil: all are commercially available. In an emergency, trial and error testing will unveil numerous ink pencils (indelible), writing inks, and stamp pad inks that will reproduce. When the ink has been applied to the master copy, do not blot. If pencil is used, be sure that the copy is strong and uniform.

   (2) When the gelatin plate is set and ready for work, sponge the plate thoroughly with cold water and allow it to set for an additional minute or two. Using a sponge, remove all excess moisture and apply the master copy, face down, on the gelatin plate. Carefully smooth the copy to ensure complete and uniform contact with prepared plate. Do not remove for at least 2 minutes. Lift one corner of the master for a gripping point and smoothly and carefully lift the master copy from the gelatin plate. The gelatin plate now bears a negative copy of the desired material and is ready to reproduce the copy.

   (3) Begin reproduction immediately after

the master copy has been removed from the gelatin plate. Reproduction is accomplished by placing a blank sheet of smooth surface paper on the gelatin plate and smoothing it into total contact by using the hand (or a rubber roller if available), then lifting the sheet from the gelatin surface. This is done as rapidly as possible to obtain as many copies as possible from one inking of the plate. One good inking of the plate may produce from 100 to 200 copies by this method, while a commercial Ditto machine may produce as many as 700 copies. In order to speed this process, one small corner of the sheet of reproduction paper is left free for gripping. This can be accomplished by permanently affixing a small piece of paper to the place on the gelatin plate where a corner of the reproduction paper would fall. This piece of paper acts as a guide and a buffer to keep that one corner of the reproduction paper from sticking. When removing the reproduction paper, lift the sheet by the loose corner; do not attempt to roll it away. The rolling action will cause the reproduction paper to curl as it dries.

(4) After completing the reproduction job, sponge the gelatin plate thoroughly with cold water and allow it to set for 48 hours or until the ink has been assimi-

lated by the gelatin. The plate is now ready to be used on a new and different job. The only way to shorten this time span is to dissolve the gelatin plate in hot water; boil off the superfluous water until the liquid is thickened to the desired consistency, and pour a new gelatin plate. Of course, two or more gelatin plates may be prepared to increase production capabilities.

APPENDIX III

AIR AND AMPHIBIOUS MESSAGES

1. Sample Drop Zone Report*

Item	Sample entry
Code name	DZ HAIRY
Location	THREE TWO TANGO PAPA TANGO SIX FOUR ONE TWO FOUR THREE
Open quadrants	OPEN ONE THREE ZERO DEG TO TWO TWO ZERO DEG AND THREE THREE ZERO DEG TO ZERO ONE TWO DEG
Recommended track	TRACK THREE SIX ZERO DEG
Obstacles	RADIO TOWER ZERO EIGHT SIX DEG SIX KM

*All items will be reported. When applicable NONE will be reported in order to preserve sequence.

2. Sample Area Drop Zone Report

Item	Sample entry
Code name	DZ JOLLY AREA
Location	PT ALFA THREE TWO TANGO PAPA TANGO SIX ONE TWO THREE FOUR FIVE PT BRAVO THREE TWO TANGO PAPA TANGO SIX ONE TWO FOUR NINE TWO
Open quadrants	NONE (NOTE: Not applicable to area DZ's, but will be reported as NONE to preserve report sequence.)
Track	TRACK THREE SIX ZERO DEG
Obstacles	PT ALFA TOWER ONE EIGHT ZERO DEG ONE ZERO KM

Item	Sample entry
Reference point	PT ALFA NEWVILLE ZERO FOUR FIVE DEG ONE FOUR KM PT BRAVO BLUE LAKE TWO ZERO FIVE DEG ONE TWO KM

3. Sample Landing Zone Report

Item	Sample entry
Code name	LZ NOBLE
Location	THREE TWO TANGO PAPA TANGO SIX ONE SIX TWO FOUR ZERO
Long axis	AXIS ONE TWO ZERO DEG
Description	FIRM SOD ONE FIVE ZERO FT BY THREE SIX ZERO ZERO FT
Open quadrants	OPEN ZERO FIVE ZERO DEG TO ONE NINE ZERO DEG AND TWO FIVE ZERO DEG TO THREE ONE ZERO DEG
Recommended track	TRACK ONE TWO ZERO DEG
Obstacles	TOWER ZERO ONE ZERO DEG FOUR KM
Reference point	OLDBURG ZERO FIVE ZERO DEG NINE KM

4. Sample Request for Airdrop or Airlanded Mission

Item	Sample entry
Code Name	DZ HAIRY PRI

Note. When requesting a mission to be flown to a drop or landing zone that has been reported to SFOB previously, it is necessary to give the code name of the DZ/LZ only. If the DZ/LZ has not been reported previously, the mission request should contain all items shown in the appropriate examples above.

Date/Time group	ZERO FIVE TWO TWO ZERO ZERO ZULU FEB
Request	ONE ZERO INDIA ALFA
Alternate	DZ HANDY ALT

Note. An alternate drop or landing zone normally will be designated whenever a mission is requested. If the alternate DZ/LZ has been reported to SFOB previously, only the code name need be given. If the DZ/LZ has not been reported previously, the mission request should contain complete information as shown in the preceding examples. Primary and alternate will always be identified as such by use of the abbreviations PRI and ALT as shown above. A date/time for the alternate will not be submitted by the requesting detachment, but will be determined by the SFOB in coordination with the air support unit, and the requesting detachment will be advised in the mission confirmation message.

5. Sample Airdrop Confirmation Message

Item	Sample entry
Code Name	DZ HAIRY PRI
Actual track	TRACK THREE SIX ZERO DEG
Actual date/time	ZERO FIVE TWO TWO ZERO ZERO ZULU FEB
Number containers or/personnel	ONE TWO PERS THREE BUNDLES
Drop altitude	EIGHT ZERO ZERO
Alternate DZ	DZ HANDY ALT
Alternate date/time	ZERO FIVE TWO TWO FOUR ZERO ZULU FEB

6. Sample Airland Confirmation Message

Item	Sample entry
Code name	LZ NOBLE PRI
Actual track	TRACK ONE TWO ZERO DEG
Actual date/time	ONE NINE ONE ZERO THREE ZERO ROMEO APR
Alternate	LZ NANCY ALT
Alternate date/time	ONE NINE ONE ONE ZERO ZERO ROMEO APR

7. Sample Beach Landing Site Report, with Mission Request

Item	Sample entry
Code name	BL WATER PRI
Location	THREE TWO TANGO PAPA TANGO ONE ONE SIX TWO THREE FOUR

386

AGO 6242C

Item	Sample entry
Description	FIRM SAND THREE ZERO ZERO M LONG BY FIVE ZERO M WIDE NO OBSTRUCTIONS
Date/time	ZERO EIGHT ZERO FOUR ZERO ZERO ZULU JUN
Request	TWO FOXTROT LIMA NINE FOXTROT MIKE EIGHT ALFA JULIETT
Alternate	BL WINDY ALT

8. Sample Beach Landing Mission Confirmation Message

Item	Sample entry
Code name	BL WATER PRI
Actual date/time	ZERO EIGHT ZERO FOUR ZERO ZERO ZULU JUN
Alternate	BL WINDY ALT
Alternate date/time	ZERO EIGHT ZERO EIGHT THREE ZERO ZULU JUN

9. Example

A sample message for any of the purposes indicated above may be constructed by simply writing the information given in the "Sample Entry" column, without breaks or paragraph headings. An example, using the samples given in paragraphs 1 and 4, is shown—

DZ HAIRY PRI THREE TWO TANGO PAPA TANGO SIX FOUR ONE TWO FOUR THREE OPEN ONE THREE ZERO DEG TO TWO TWO ZERO DEG AND THREE THREE ZERO DEG TO ZERO ONE TWO DEG TRACK THREE SIX ZERO DEG RADIO TOWER ZERO EIGHT SIX DEG SIX KM WHITE LAKE ONE SEVEN TWO DEG SEVEN KM ZERO FIVE TWO TWO

AGO 6242C

ZERO ZERO ZULU FEB ONE ZERO INDIA
ALFA DZ HANDY ALT

APPENDIX IV

FIELD EXERCISE

Section 1. INTRODUCTION

1. General

a. This appendix outlines a field exercise for both headquarters and operational elements of a Special Forces group which have attained the advanced unit level of training. The exercise is a joint maneuver and incorporates other service support of unconventional warfare activities to include participation by the Joint Unconventional Warfare Task Force (JUWTF) in overseas theaters. The exercise may be modified to meet local requirements; however, to retain maximum realism, the basic structure should not be altered.

b. The exercise consists of two phases—a preparatory phase, involving Special Forces participants, of 30 to 45 days; and the exercise proper of 30-days' duration.

2. Purpose and Scope

a. The exercise is a training vehicle designed to—

(1) Train the headquarters and operational elements of the Special Forces group.

(2) Provide for joint training between the Special Forces group and support elements from other services.

(3) Exercise communications systems over extended distances.

(4) Test, in overseas theaters, certain aspects of unconventional warfare plans such as—

　(*a*) The logistical support plans.

　(*b*) Activation of the SFOB.

　(*c*) Command relationships between the Special Forces group, the JUWTF, and other unified command elements.

b. In addition to the unconventional warfare objectives of the exercise, conventional troop participants benefit as follows:

(1) *Aggressor force.* Training in counterguerrilla operations with stress on small-unit actions. Most aggressor activity involves small units on semi-independent missions transported and supplied largely by aircraft.

(2) *Guerrilla troops.* The troops participating as the guerrilla rank and file are trained by the Special Forces operational detachments. Training stresses small-unit operations conducted at night in enemy rear areas. Demolitions, air resupply procedures, raids, and ambushes receive heavy emphasis. The ex-

ercise is an excellent training vehicle for small-unit leaders.

c. The exercise is a joint field maneuver commencing with the establishment of the SFOB; continuing with initiation of operations; and ending with demobilization of the area command.

3. Conduct of the Exercise

a. The exercise may be conducted in a single cycle or extended over two or more cycles. Each cycle is a complete exercise (30 days) for all participants.

    (1) *Single cycle exercise.* This exercise requires commitment of the entire Special Forces group. The single cycle method has the advantage of involving for a relatively short period of time the Special Forces group and large numbers of other participants. The major disadvantages of the single cycle method is that all detachments of the group are not able to participate in an operational role. The same number of detachments is required for guerrilla cadre duties as for operational units. Additional personnel in excess of the headquarters and headquarters company are needed to staff the exercise control group.

    (2) *Two or more cycles.* This method has several advantages. Since all detachments usually have not achieved the same level of training, the less qualified

detachments participate in later cycles when they reach the necessary level of readiness. All detachments are employed in the operational role. Those detachments completing cycle one as operational detachments may be used during cycle two in their same area as the guerrilla cadre detachment. The major disadvantage is the large number of supporting elements required over a relatively long period of time.

b. The exercise should be conducted annually by detachments that have reached advanced unit training. The exercise is designed to be the high point of the Special Forces group training year.

c. Realism throughout the exercise is stressed. The commander responsible for its preparation prescribes restrictions to meet safety requirements. Within these limits, the director insures that realistic procedures are incorporated.

d. Maximum latitude of action is allowed all participants in the conduct of the exercise within necessary maneuver restrictions.

4. Participants

a. Friendly Forces.

(1) Special Forces group.

 (*a*) Headquarters and headquarters company establishes the SFOB. Signal company provides base radio communication within the SFOB and with

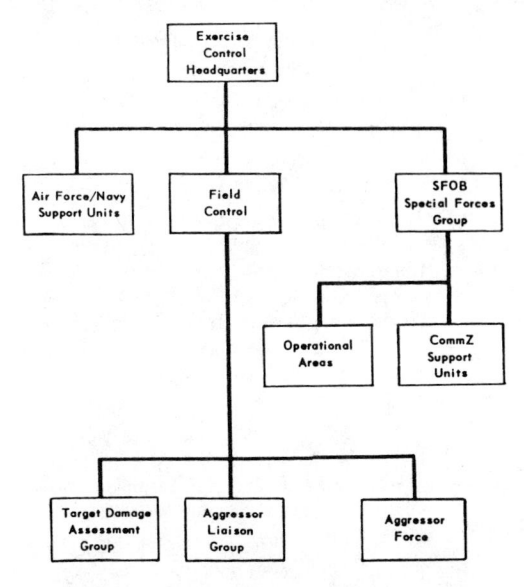

Figure 87. Organization for the field training exercise.

        SF operational detachment, and other central communications as required.
- (b) Special Forces companies.
    1. Provide operational detachments.
    2. Provide detachments to cadre the guerrilla force on the basis of one guerrilla cadre detachment per operational sector. The guerrilla cadre detachment should contain all of its regularly assigned personnel with

the possible exception of communications personnel.

(2) United States Air Force provides air support units as required.

(3) United States Navy provides naval support as required.

(4) JUWTF provides operational control (control headquarters) for designated theater unconventional warfare forces. If the exercise originates from the continental United States (CONUS), the JUWTF may be played by a Special Forces group operational detachment C.

(5) Theater Army Communications Zone (CommZ) furnishes administrative and logistical support. If the exercise originates from CONUS, this support is provided by the appropriate continental army commander.

(6) Theater or CONUS intelligence agencies provide intelligence support.

(7) Theater Army, or in CONUS, the continental army area furnishes troops for the guerrilla force. A minimum of 50 non-Special Forces guerrillas are provided for each Special Forces guerrilla cadre detachment.

(8) Personnel to participate as evaders in evasion and escape, should be aircrew members and furnished from Air Force, naval air, or Army aviation units.

b. Aggressor Forces. The size of the aggressor force depends upon the size of the exercise area and the strength of the deployed unconventional warfare force in the areas. As a general rule, the aggressor to friendly force ratio should be no less than 3 to 1 in favor of aggressor. An exercise area containing 400 to 500 guerrillas and Special Forces troops requires an aggressor force equal to 2 or more infantry battalions. The aggressor force should contain the following additional elements:

(1) Elements of one airmobile battalion as required.

(2) One troop of an armored cavalry squadron.

(3) Military intelligence elements—
 (*a*) OB specialists.
 (*b*) Interrogators.
 (*c*) Intelligence corps (INTC) personnel.

(4) Army aviation elements other than the airmobile battalion.

(5) Communications security or communications intelligence elements.

(6) Normal administrative and logistical support elements for a brigade.

(7) Elements of a psychological operations company (loudspeaker and leaflet).

(8) Appropriate aerial surveillance and target acquisition elements with supporting military intelligence imagery interpreters.

5. Exercise Area

a. Location. The exercise area should contain—

(1) Terrain of varying ruggedness to facilitate development of a guerrilla force.

(2) Suitable "live" targets (railroads, highways, and telecommunications systems) to be used for actual and simulated attack.

b. Size. The area must be large enough to permit logical subdivision into sufficient operational areas or sectors for the number of detachments to be deployed. A sector of roughly 700 square miles is adequate for a single detachment in an exercise of 30-days duration. Size is further influenced by the nature of the terrain.

c. Organization. The exercise area is subdivided into guerrilla warfare operational areas (GWOA) and sectors. As a general rule, one operational detachment may organize each sector. Detachments B and C are either superimposed upon an A detachment or employed in separate areas.

d. Relationship of Guerrilla Warfare Operational Areas (fig. 81). Operational areas and sectors are ideally located in a circular formation around the field control headquarters and adjacent to each other. Although this arrangement is not always an operational ideal, it facilitates administrative control by the exercise director and allows the employment of more operational detachments. It also tends to increase the effectiveness of the aggressor force.

6. Planning and Preparatory Phase

a. Planning. Planning the FTX requires 6 to 9 months, depending upon the command structure in the theater, the number and types of participants, and their relative geographic locations. The headquarters directing the exercise appoints planning personnel (the exercise director and necessary staff assistance) far enough in advance of the exercise to insure adequate preparatory planning and coordination by all participants.

b. Special Forces Group.

  (1) *Preparation.* Special Forces troops should be given 30 to 45 days for preparation. This preparatory phase is designed to:

  (a) Permit operational detachments to reach a high peak of unit proficiency before the exercise.

  (b) Teach special techniques to be employed or tested during the exercise.

  (c) Train the operational detachments in conjunction with the supporting air or naval units.

  (d) Train guerrilla cadre detachments.

  (e) Allow the guerrilla cadre detachments to reconnoiter and prepare their operational areas.

- (f) Conduct a CPX for headquarters elements.
- (g) Train the exercise control personnel.
- (2) *Facilities.* If the SFOB is not physically established, the headquarters elements of the Special Forces group and theater support troops erect the necessary facilities during the preparatory phase.

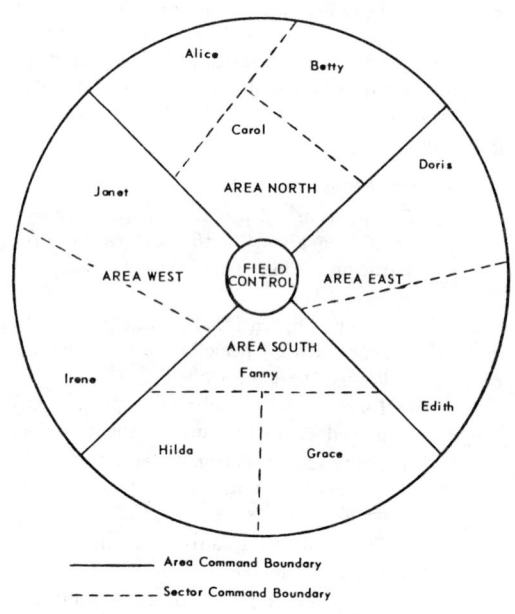

Figure 88. The exercise area.

(3) *Programing.* One week is programed between the end of the preparatory training and the commencement of the exercise. This period allows for maintenance of equipment and final preparation.

(4) *Operational detachments.*
  (a) The operational detachments undergo an intensive period of training to prepare for the exercise. This training is conducted as a series of short (4 to 5 days) field exercises which incorporate the following:
   1. Air or amphibious operations in conjunction with supporting air or naval units.
   2. Reconnaissance and selection of guerrilla bases and other related areas (evasion and escape contact areas, mission support sites, and caches).
   3. Tactical operations.
   4. Communications.
   5. Organization and employment of area command forces.
   6. Training of guerrilla forces.
   7. Field craft and cross-country movement.
   8. Evasion and escape identification techniques.
  (b) In addition, operational detachments engage in certain preparatory training not conducted in the field, such as:

1. Area studies of the entire exercise area.
   2. Demolitions simulation procedures.
   3. Self-aid.
   4. Administrative briefings.
   5. Foreign weapons.
(5) *Guerrilla cadre detachments.*
   (a) Guerrilla cadre detachments are Special Forces operational detachments that provide the guerrilla force with its leadership and also serve as a part of the exercise control system. The success of the exercise as a training vehicle rests to a great extent upon the efforts of the guerrilla cadre detachment. These detachments are selected because of their experience and advanced state of training. Where possible, the guerrilla cadre detachment has already participated as an operational detachment in a similar exercise.
   (b) The guerrilla cadre detachments undergo short formal training (3 to 5 days) conducted by the control group. This training includes—
   1. Administrative briefings.
   2. Area studies of assigned operational area. For this area study the guerrilla cadre detachment prepares a terrain analysis of their area.
   3. Requirements to be presented to operational detachments. These are the

requirements to be injected into the exercises on a predetermined time schedule.

4. Role of the guerrilla cadre detachment in the exercise control system.

(c) The majority of time in the preparatory phase is utilized by the guerrilla cadre detachments for a physical reconnaissance of the operational areas. This reconnaissance should take at least 3 weeks. It permits the detachments to become thoroughly familiar with their areas. The cadres establish contacts with civilians in the area and organize the nucleus of auxiliary units. They attempt to elicit support from the civilians for the friendly forces. They locate possible guerrilla bases, mission support sites, evasion and escape contact areas, caches, and other installations. They become familiar with the important targets in the area. They study the terrain and suitable routes for movement. The results of this reconnaissance are reported to the exercise director. This report includes as a minimum—

1. An infiltration DZ and alternate DZ.
2. Suitable landing zones for aircraft, or water landing sites.
3. Recommended routes for ambush problems.

4. Alert sites for emergency pickup of the participants.
5. Background stories on the key resistance leader in the area (guerrilla cadre detachment personnel).
6. Administrative contact points for meeting exercise control personnel and visitors.

(6) *Headquarters and headquarters company, and signal company.* The headquarters and headquarters company, and signal company participates in a short CPX (2 to 5 days) during the preparatory phase. The CPX enables the Special Forces group commander to test his staff and communications elements. Operational detachment radio personnel may be included in the SFOB communications net during this CPX. The CPX may be conducted in conjunction with one of the FTX's for operational detachments (6b(4) above).

(7) *Exercise control group.*
  (a) The exercise control group uses the preparatory phase to train its various elements. These elements are—
    1. Control headquarters.
    2. Target damage-assessment group (TDAG).
    3. Guerrilla cadre detachments.
    4. Aggressor liaison group (ALG).
  (b) Control headquarters trains the necessary operations and clerical personnel

    to handle field control and control headquarters. This consists of primarily on-the-job training.
- (*c*) TDAG trains its personnel in assessing damage and demolitions simulation techniques for the exercise. TDAG personnel reconnoiter the exercise area to familiarize themselves with the road net and target systems.
- (*d*) ALG reconnoiters the area to familiarize themselves with the terrain and road net. In connection with this reconnaissance ALG arranges for aggressor campsites and field landing zones.

c. Other Services. The Air Force or Navy support elements participate in joint exercises with elements of the Special Forces group during the preparatory phase. Final planning by these elements is based upon briefings by the Special Forces group and the JUWTF.

d. Aggressor Force. The ALG assists the aggressor force commander to prepare for the exercise. The aggressor force conducts area studies of the exercise area and receives administrative briefings. Instruction in counterinsurgency operations is presented to all personnel by the ALG. Aggressor reconnoiters the area; however, care must be exercised that aggressor reconnaissance parties do not interfere with guerrilla cadre detachments.

Section II. OPERATIONS PLAN

7. Situation

a. The general situation used as a background for the exercise may take one of several forms. Some examples are—

   (1) A general war with either early or delayed deployment of operational detachments.

   (2) A limited war with early or delayed deployment of operational detachments.

   (3) The use of nuclear weapons by either or both sides may be included in the general situation.

b. The following assumptions are incorporated into the special situation:

   (1) A resistance potential exists.

   (2) The resistance potential will accept U.S. sponsorship.

   (3) Initial contact with resistance elements has been established.

   (4) The situation favors employment of Special Forces detachments.

8. Scenario

a. General. The exercise is further subdivided into a series of six phases. Each of these phases consists of a number of requirements, some accomplished at or initiated by the SFOB, and the remainder originating from the field.

b. Sequence. For sequence of events, see figure 89.

c. Phase One. The headquarters and headquarters company and operational detachments are ordered to move from home station to the SFOB. If in an overseas theater the movement conforms as closely as possible to an actual alert move. The guerrilla force is deployed into the exercise area during phase one.

d. Phase Two. The Special Forces group occupies and activates the SFOB. Operational detachments are isolated in the briefing center and prepare for deployment.

e. Phase Three. Special Forces detachments infiltrate designated guerrilla warfare operational areas.

f. Phase Four. Operations to support the unified commander including—

    (1) Automatic, emergency, and on-call resupply.

    (2) Organization and development of independent sector commands.

    (3) Training and employment of resistance elements to include conduct of ambushes and raids, interdiction techniques, intelligence operations, tactical cover and deception techniques, and psychological warfare operations.

    (4) Evasion and escape operations.

    (5) Organization of area commands composed of a B and C detachment and subordinate operational detachments.

g. Phase Five. Operations include—
 (1) Resupply.
 (2) A coordinated interdiction mission. These missions are initiated by the SFOB but coordinated by the B or C detachments.
 (3) Evasion and escape operations.
 (4) Transfer control of operational areas from SFOB to field army, or lower echelon.
 (5) Establishment of physical liaison between area commands and field army elements.
 (6) Integrated CBR warfare requirements.
 (7) Operations to assist field army units. This operation may be the seizure of a key installation or critical terrain or assistance to an airborne, air landed, or amphibious operation.
 (8) Linkup with conventional forces.

h. Phase Six. Post-linkup operations—the demobilization of the guerrilla force.

i. Additional Information. For a further discussion of operations, see section VII of this appendix.

Section III. INTELLIGENCE PLAN

9. General

The intelligence plan is implemented in three ways—first, by actual intelligence developed in the operational area; the principal means of providing this intelligence are the aggressor force and the guerrilla commander; second, simulated intelligence information provided by the guerrilla commander; and third, intelligence provided by the SFOB.

10. Aggressor

a. The actions of the aggressor force provide intelligence play. The area command is required to develop all possible intelligence of the aggressor force.

b. Specific items of aggressor information are reported to the SFOB. These items are restricted to significant aggressor activity.

c. For aggressor sequence of events, see figure 90.

11. Guerrilla Commander

a. As a result of his reconnaissance in the preparatory phase, the guerrilla commander is in possession of much intelligence data about the operational area, particularly with regard to—

   (1) Potential assets.
   (2) Target systems.
   (3) The terrain.
   (4) Unfriendly civilians.

b. This intelligence data is furnished to the operational detachment commander as required for the conduct of operations.

12. SFOB

a. During the preparatory phase, operational detachments have accomplished their general area study based upon material provided by the group S2. In the briefing center of the SFOB, detachments are given operational area intelligence. This intelligence includes—

(1) Latest aggressor order of battle.
(2) Instructions for contacting resistance elements.
(3) Intelligence to support the infiltration plan.

b. SFOB may attach selected individuals to the detachments in the briefing center for the intelligence play. This individual furnishes the detachment with first-hand knowledge of the area to include information about key personnel. Such personnel may infiltrate with and accompany the detachment during deployment.

c. Once deployed, the SFOB continues to furnish intelligence to the area command as appropriate.

d. The SFOB prescribes those items of intelligence information to be reported by area commands. Reporting requirements are outlined in the intelligence annex and SOP. Examples of intelligence data required by the SFOB are—

(1) Identity and location of major aggressor units (battalions, regiments, and principal headquarters).
(2) Location of missile sites.

(3) Location of air defense systems.

(4) Major aggressor activity such as movement into and withdrawal from the operational area and large scale counterguerrilla activities.

e. Brevity codes are employed to transmit required intelligence data to SFOB.

Section IV. ADMINISTRATIVE PLAN

13. General

The administrative plan of the exercise includes all actual administrative and logistical support necessary for the participants. The Special Forces group provides for the logistical and administrative support of its own troops and the guerrilla force using established procedures. Other participating services furnish their own logistical and administrative support. The aggressor force is supported by its headquarters. In overseas theaters the JUWTF may coordinate administrative and logistical support plans among the participating services. This section discusses administrative and logistical support peculiar only to the operational areas.

14. The SFOB

The SFOB fulfills its normal role by providing administrative and logistical support for elements located at the base and for the area commands.

SUNDAY	MONDAY	TUESDAY	WEDNESDAY	THURSDAY	FRIDAY	SATURDAY
	1 Movement to the SFOB. Note 1. ←PHASE ONE→	2 Guerrilla Force Deployed.	3 Occupation and activation of SFOB. Note 2.	4 ←PHASE TWO→	5 Preparation for Deployment of operational detachments. Note 3. Primary Deployment Date.	6 ←PHASE THREE
7 Alternate deployment date. ←PHASE THREE→	8 Automatic resupply done.	9	10 Notes 4 through 8.	11	12	13 ────PHASE FOUR
14	15	16 Deployment of additional command detachments. Note 9.	17	18	19 Activation of area commands. Note 10	20 ────PHASE FIVE
21 PHASE FOUR───	22	23 Coordinated interdiction mission. Note 11	24 ←PHASE FIVE─	25	26 Transfer of operational control to field army. Note 12. Integrated CBR Problem. Note 15.	27 Field army element/liaison party joins area command. Note 13.
28 Direct support mission of field army element. Note 14. Integrated CBR Problem. Note 15. ←PHASE FIVE─	29	30 Demobilization. Note 16. ←PHASE SIX→				

Figure 89. Operations calendar.

Note 1. This is a tactical move. In the event this phase is accomplished in less than two days, the additional time is utilized in phase two. The guerrillas are received, processed, and moved to their areas during this period.

Note 2. SFOB is already physically established.

Note 3. Two days are allowed for deployment, one a primary date, the second an alternate. Some C and B detachments may be deployed along with the A detachments. Infiltration is accomplished during the hours of darkness.

Note 4. Each sector is required to conduct a minimum of two ambushes during phase four.

Note 5. Each sector is required to conduct one raid against a well-guarded target, during phase four.

Note 6. Interdiction operations conducted at the option of the sector command during phase four.

Note 7. Resupply operations continued. These operations are initiated by the sector commands. Emergency resupply accomplished if required.

Note 8. Evasion and escape action commences with the random introduction of evaders throughout the exercise area. Evasion and escape continues into phase five.

Note 9. B and C detachments scheduled for subsequent deployment are infiltrated.

Note 10. Area commands are activated. Each area command consists of one B or C detachment (area) and one or more subordinate detachments (sectors).

Note 11. The coordinated interdiction mission is included in detachment operation order for execution when directed by the SFOB.

Note 12. Transfer of operational control to field army is completed by SFOB.

Note 13. A field army liaison party consisting of an officer and communications personnel is paradropped or air landed in the operational area.

Note 14. Missions to support conventional forces may be one or a combination of the following:

a. Protection of a key installation.
b. Seizure of critical terrain.
c. Support of:

  (1) Airborne/air landed operations.
  (2) Amphibious operations.

Note 15. A CBR warfare requirement is integrated into phase five during execution of the mission to support conventional forces.

Note 16.

a. Demobilization instructions sent to area commands via field army liaison party. Guerrilla commanders confront special forces detachments with various demobilization problems.

b. Guerrillas are moved to a base camp where appropriate demobilization ceremonies are conducted.

Figure 89—Continued.

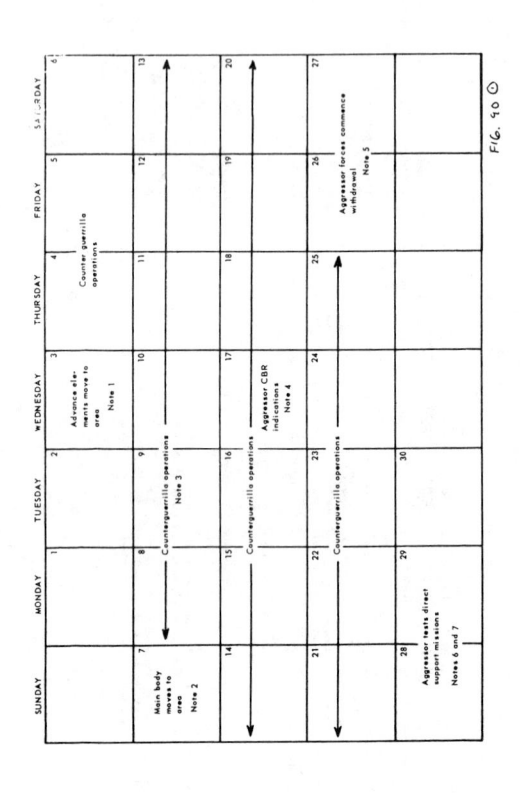

Figure 90. Aggressor calendar.

Note 1. Advance elements include:
 a. Command element.
 b. Reconnaissance element.
 c. Military intelligence units. CIC personnel occupy the exercise areas from one to two weeks in advance of the exercise commencement date.
 d. ASA units.
 e. Army aviation elements.
 f. Psychological warfare units.
 g. Aggressor liaison group.

Note 2. Main body consists of the infantry battle group and administrative and logistical support units. Elements of the aggressor liaison group join the main body.

Note 3. Aggressor force provides guards for selected key installations and simulated targets such as ammunition depots and missile launching sites. Aggressor plans and executes his own counterguerrilla operations advised by the aggressor liaison group. If aggressor tends to be inactive, the exercise director orders search operations of "quiet" areas. He avoids pinpointing guerrilla forces but selects likely areas for counter-guerrilla operations.

Note 4. Aggressor CBR warfare indications commence and continue throughout the exercise. Aggressor troops carry protective masks and engage in frequent CBR warfare drills.

Note 5. Roughly one-half of the aggressor force withdraws during this period. These troops then change roles to that of conventional U.S. forces. As conventional forces they effect link-up with the guerrilla units. They may execute airborne/air landed or amphibious operations or move to guerrilla protected installations or terrain.

Note 6. Aggressor forces test the guerrilla execution of the missions to support conventional U.S. forces. They attack guerrilla defended installations or terrain and attempt to disrupt airborne/air landed or amphibious operations. Aggressor employs CBR training agents against the guerrillas and U.S. Link-up forces.

Note 7. As aggressor forces complete the test of guerrilla missions to support conventional U.S. forces, they withdraw from the exercise area. Aggressor forces which reverted to the U.S. role withdraw after physical link-up with the guerrillas is completed.

Figure 90—Continued.

15. External Supply

The SFOB provides supplies using the delivery means available. A catalogue supply system facilitates requesting procedures. Supplies consist of—

a. Light machine guns and foreign weapons.

b. Blank ammunition, pyrotechnics, smoke, and other training ammunition simulators.

c. Signal equipment such as extra radios.

d. Basic medical supplies such as bandages, APC's, cough medicine, insect repellent, and foot powder.

e. Demolitions material.

f. About 40 percent of the total ration requirement for area commands is provided by the SFOB. Although this ration requirement may seem unrealistic, the rations are used to represent other types of supplies such as ammunition and demolitions, the issue of which, in normal amounts, is not feasible.

16. Air Support Plan

The air support plans incorporate the following elements—

a. Each mission has an alternate date/time 24 hours after the primary date/time in the event of air or ground abort of the mission. Aborts are not rescheduled beyond 24 hours unless unlimited air support is available.

b. Automatic, emergency, and on-call resupply missions are planned for each operational detachment. These missions are executed as required.

c. Except for automatic and emergency resupply missions, the operational detachment retains the initiative to request resupply operations. The air support unit may schedule blocks of time for on-call resupply missions in each area. This procedure facilitates proper flight planning and maximizes flying safety. Each area plans its resupply operations within their allotted time blocks. These time blocks are staggered so as not to present a pattern of operations. Generally, resupply missions are flown during the hours of darkness; however, some dawn or dusk missions may be desirable, particularly in remote areas.

17. Internal Supply

a. Guerrilla cadre detachments cache approximately 60 percent of the total ration requirement in the operational area before commencement of the exercise. These rations are controlled by the guerrilla commander and represent various methods of internal procurement such as levy, barter, or confiscation. They also serve to give the guerrilla force a degree of logistical independence and preclude a degeneration of the maneuver into a survival exercise.

b. Additional rations, weapons, demolitions material, blank ammunition, CBR protective equipment, and other supplies are provided by means of ambushes and raids. These methods are valuable as a backup to air or water resupply plans in the event of prolonged inclement weather. In all cases, aggressor provides a guard detail or escort for supplies introduced in this fashion.

18. Transportation

As a general rule, the use of mechanical means of transportation in operational areas is forbidden before linkup, except in those instances where aggressor transportation is captured. In order for the exercise to be realistic, guerrilla forces move men and material by foot or by using locally procured animals. Once linkup is completed, restrictions on the use of mechanical transportation are lifted.

19. Hospitalization and Evacuation

a. Minor injuries and sickness are treated in operational areas by detachment medical personnel.

b. A helicopter is provided on a standby basis at the field control headquarters for evacuation of injured personnel.

c. The group surgeon delineates sickness and injury to be treated in the field without evacuation.

20. Personnel Procedures

Each operational detachment undergoes an administrative processing before deployment. Once deployed, personnel procedures are handled according to SOP. The SOP minimizes personnel action required on exercise participants in the field.

21. Miscellaneous

a. Nonexpendable items (parachutes, air delivery containers, or damaged equipment) are recovered administratively by TDAG personnel. This recovery is coordinated through the guerrilla commander and accomplished in such a manner as to preclude compromise of the guerrilla force.

b. The administrative infiltration and removal of visitors and other non-operational personnel is accomplished as cited above. Care must be exercised by field control not to use helicopters and other compromising means of transportation for these activities.

c. The administrative infiltration and evacuation of operational personnel should be kept to an absolute minimum. When infiltration or evacuation, such as the return of prisoners to operational areas, is necessary, it is accomplished as cited in *a* above.

d. Civilian clothing is authorized for use by selected members of guerrilla cadre detachments.

Section V. CONTROL PLAN

22. General

The control elements for the Special Forces field exercise are—

a. Control headquarters.

   b. Field control.

   c. Target damage assessment group (TDAG).

   d. Aggressor liaison group (ALG).

   e. The guerrilla commanders.

23. Control Headquarters

a. The control headquarters consists of the exercise director and his staff. In overseas theaters, the JUWTF may provide the control headquarters. In CONUS, the control headquarters may be a C detachment, or, if activated, the augmentation detachment. Normally, the control headquarters is located at the SFOB.

b. The control headquarters supervises the preparation of the exercise, to include coordination with all participants.

c. Control headquarters plans and supervises the preparatory training phase for—

   (1) The guerrilla cadre detachments.

   (2) ALG.

   (3) Field control.

   (4) The aggressor force.

d. Control headquarters may plan and supervise the preparatory training for the operational detachments.

24. Field Control

a. This element functions as the control headquarters agency in the exercise area. Field control is an administrative element only and as such is responsible for—

    (1) Coordination of activities of the aggressor force.

    (2) Supervision of TDAG.

    (3) Administrative control of guerrilla commanders.

    (4) Briefing and handling of visitors to the exercise area.

    (5) Arrangements for raid targets, ambushes, and the infiltration and evacuation of visitors, evaders, and other participants.

    (6) Emergency evacuation of injured or deceased personnel.

b. Field control personnel train for 1 week to enable them to perform their duties.

c. Field control personnel have the status of umpires in ruling on the outcome of contact actions between friendly elements and aggressor.

25. Target Damage Assessment Group (TDAG)

a. TDAG operates under the field control headquarters (fig. 92).

b. TDAG consists of specially trained Special Forces operational personnel who assess reported guerrilla damage. Targets to be attacked are reported to field control who then direct a TDAG team to the location to assess simulated damage.

c. In addition to target assessment, TDAG may perform the following tasks:

   (1) Ground infiltration and evacuation of personnel.
   (2) Operation of a holding area for personnel who have been temporarily removed from the exercise.
   (3) Provide control personnel to accompany convoys which are to be ambushed.
   (4) Recovery of nonexpendable equipment such as parachutes and delivery containers.

d. TDAG personnel receive 1 to 2 weeks training during the preparatory phase. This training consists of—

   (1) Reconnaissance of the road and target system in the exercise area.
   (2) Techniques of damage assessment and target simulation.
   (3) Administrative briefings.
   (4) Procedures for umpiring ambushes and raids.

e. TDAG personnel have the status of umpires in ruling on the outcome of contact actions between friendly elements and aggressor.

26. Aggressor Liaison Group

a. ALG consists of selected officers and senior noncommissioned officers who are the field control element with the aggressor force (fig. 93).

b. ALG is responsible for—

   (1) Preparation and conduct of counterguerrilla instruction for the aggressor force during the preparatory phase.

   (2) Assisting the aggressor force in its pre-exercise reconnaissance of the area.

   (3) Assisting the aggressor force in the procurement of bivouac sites, landing zones, and other installations in the maneuver area.

   (4) Advising the aggressor force in counterguerrilla operations.

27. The Guerrilla Commander

   a. Guerrilla commanders probably are the most important single control element in the exercise. Because of their location with the friendly forces, they are usually present during operations where contact between guerrilla and aggressor is most likely.

   b. The guerrilla commander prepares that portion of the exercise pertaining to the resistance potential in his area. This consists primarily of background stories for the guerrilla cadre detachment personnel and the guerrilla force.

   c. He selects the primary and alternate infiltration DZ's or water landing infiltration site.

   d. He orients and organizes his guerrilla force.

   e. He implements portions of the scenario by presenting situations and requirements to the

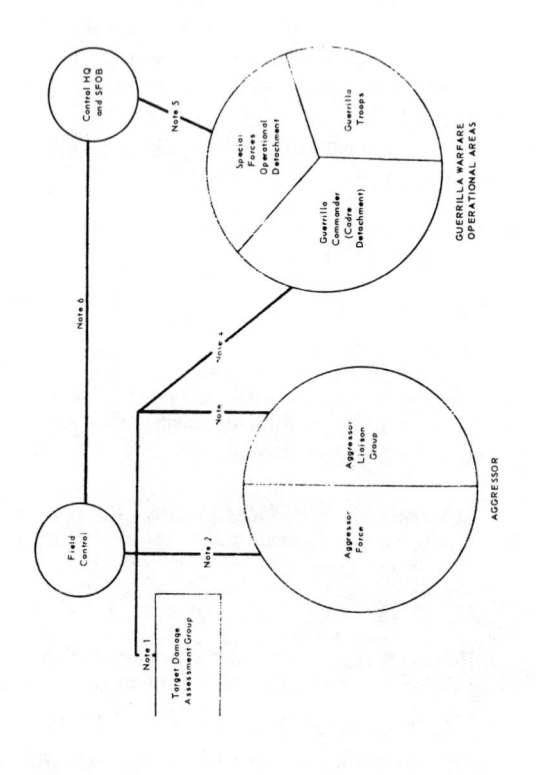

Figure 91. Exercise control system.

Note 1. Field control directs activities of TDAG based upon information furnished by guerrilla cadre detachment.

Note 2. Field control acts as aggressor higher headquarters and controls aggressor activities. Daily joint briefings are conducted by both field control and aggressor. These briefings provide field control with a means of monitoring aggressor operations.

Note 3. Field control maintains contact with aggressor liaison group to monitor aggressor operations. Instructions to aggressor are issued directly to the aggressor commander, not through the aggressor liaison group.

Note 4. Field control coordinates with the guerrilla commander for certain tactical play (raid targets, ambushes) and administrative matters concerning the guerrilla force are handled through this channel.

Note 5. SFOB directs the operational activities of the guerrilla force through the special forces operational detachment. The external resupply plan is also controlled through this channel.

Note 6. Control headquarters maintains close contact with field control. This contact enables control headquarters to keep abreast of the situation in the field.

Figure 91—Continued.

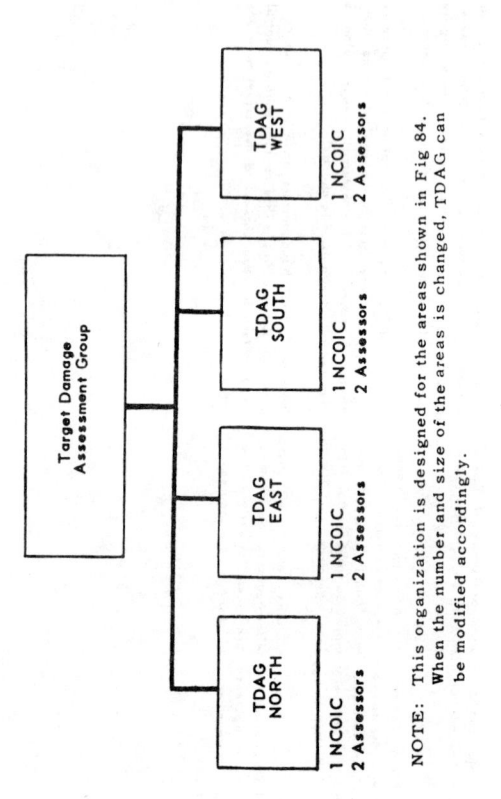

```
                    ┌──────────────┬──────────────┬──────────────┐
   Target Damage    │              │              │              │
   Assessment Group │              │              │              │
                    │              │              │              │
              TDAG NORTH     TDAG EAST      TDAG SOUTH     TDAG WEST
              1 NCOIC        1 NCOIC        1 NCOIC        1 NCOIC
              2 Assessors    2 Assessors    2 Assessors    2 Assessors
```

NOTE: This organization is designed for the areas shown in Fig 84. When the number and size of the areas is changed, TDAG can be modified accordingly.

Figure 92. Functional chart, TDAG.

operational detachment on a prearranged schedule. He coordinates with the field control headquarters daily and keeps them informed of the situation in the sector. He arranges convoys for ambush training and other special targets, such as missile launching sites, or radar installations. He coordinates the administrative evacuation of personnel and nonexpendable equipment.

f. He has the status of an umpire in ruling on the outcome of contact between friendly elements and aggressor.

28. Control Notes

a. The Special Forces field exercise is not as closely controlled as are conventional maneuvers. Since contact between opposing sides is not as frequent, a less rigid umpire system is used. Umpires are not required for each unit, but are assigned to those units most likely to effect contact. The exercise director has at his disposal sufficient means of control to umpire those actions where contact between opposing forces is most probable, such as raids, ambushes, and missions to support conventional forces.

b. The exercise director controls the exercise using the established control system. He insures that all control elements are fully informed on the situation as it pertains to them. Through prior planning and close liaison with guerrilla commanders and the aggressor liaison group, he is able generally to anticipate the action of the exercise.

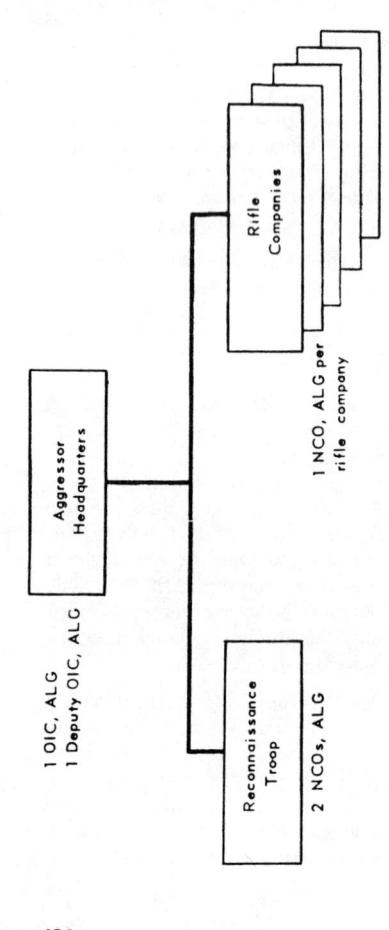

```
                    ┌──────────────┐
                    │  Aggressor   │
                    │ Headquarters │
                    └──────┬───────┘
        ┌──────────────────┴──────────────────┐
┌───────────────┐                     ┌──────────────┐
│Reconnaissance │                     │    Rifle     │
│    Troop      │                     │  Companies   │
└───────────────┘                     └──────────────┘
```

1 OIC, ALG
1 Deputy OIC, ALG

2 NCOs, ALG

1 NCO, ALG per rifle company

NOTE: This organization is designed for the areas shown in Figure 84. When the number and size of the areas is changed, TDAG can be modified accordingly.

Figure 93. Distribution of aggressor liaison group.

c. Control personnel, except guerrilla commanders, wear distinctive markings. All control personnel carry special identification cards and are exempt from capture by the other participants.

29. Directives

a. The following separate directives assist the exercise director in preparing units for and exercising control over the exercise:

    (1) A general memorandum, including the purpose, scope, objectives, participants, and organization for the exercise.

    (2) An aggressor memorandum.

    (3) A memorandum for friendly forces. This may be issued in the form of an operational SOP.

    (4) A memorandum for the guerrilla cadre detachments.

    (5) A memorandum for the units providing the rank and file of the guerrilla force.

    (6) Operation orders for friendly forces and aggressor.

b. These directives are prepared and issued separately to the various participants because each category of participants requires different operational instructions. Since the administrative instructions are generally the same for all participants, they are included in the memoranda for each of the participants.

Section VI. ORIENTATION AND CRITIQUE PLAN

30. Orientation

a. Briefings are conducted well in advance of the exercise to allow participants sufficient time for planning and preparation. Because the exercise is joint in nature and involves troop units not under the control of the Special Forces group, at least 6 to 9 months is usually required for planning.

b. Separate briefings are conducted by the exercise director for the following:

(1) Special Forces operational detachments.
(2) Special Forces headquarters and headquarters company and signal company.
(3) Aggressor commanders and staffs.
(4) Guerrilla cadre detachments.
(5) Units furnished guerrilla troops.
(6) Air Force and Navy elements.
(7) Control personnel.
(8) Other interested headquarters.

31. Critique

a. Separate critiques are prepared and implemented by the exercise director for the following:

(1) Special Forces operational elements and guerrilla cadre detachments.
(2) Special Forces support elements.
(3) Aggressor units.

(4) Guerrilla troops.

(5) Air Force and Navy elements.

b. Control personnel attend critiques as appropriate.

c. Special Forces operational detachments and guerrilla cadre detachments are critiqued together by sector. These critiques include other services participants, the Special Forces group commander and staff, and selected control personnel. Both the operational detachment and the guerrilla cadre detachment are given the opportunity to outline their actions during the exercise. They are followed by the group staff, other service representatives, and the exercise director. Four hours per sector should be allowed for this critique. Critiques are scheduled to commence 2 or 3 days after termination of the exercise to allow participants time for preparation.

d. Aggressor critiques are scheduled at the convenience of the aggressor force. Again time is allowed for the exercise director to assess results and prepare the critique.

e. Air Force and Navy commanders and staffs may be included in the critique for Special Forces support elements (SFOB) or critiqued separately. Supporting services furnish representatives for the operational detachment critique.

f. Guerrilla troops are critiqued during or immediately following the demobilization phase. As a part of their critique, guerrilla troops may be tested on training received from the operational detachments.

g. The Special forces support elements (SFOB) are critiqued separately from the operational elements. Comments by operational detachments regarding the SFOB are included in this critique when appropriate.

Section VII. SUGGESTIONS FOR THE EXERCISE DIRECTOR

32. General

a. This section is designed to assist the exercise director and his staff in the preparation and conduct of the Special Forces field exercise described in preceding sections of this appendix. It deals with those features of the Special Forces field exercise which, when applied with imagination, enhance its value as a training vehicle for all participants.

b. Although the primary purpose of the exercise is to train Special Forces troops, their supporting elements, and Air Force and Navy elements, the exercise director should not slight the training value which the exercise holds for other participants, particularly the aggressor force and troops used as guerrillas. Proper integration of aggressor and guerrilla troops into the exercise will serve to improve the combat readiness of these units and promote among conventional unit commanders a desire to participate in future Special Forces exercises.

33. Responsibilities of the Exercise Director

a. Supervises the preparation and conduct of the exercise.

b. Advises the commander directing the exercise on personnel requirements. The most qualified and experienced personnel should be assigned the task of preparation. Supporting Air Force and Navy personnel assist in planning their portions of the exercise.

c. Coordinate all planning action required by the participants. He is informed of the status of planning and preparation of the participants. In those instances where participants are lagging in their preparations, he informs the commander directing the exercise. The exercise director is prepared to offer contingency solutions if participation of a major element is curtailed or withdrawn from the exercise.

d. During the exercise proper the director normally is located at the control headquarters with the SFOB. His principal deputy is located at the field control headquarters.

34. Selection and Assignment of Detachments

a. General. Detachments are designated as "Special Forces operational" or "guerrilla cadre" as far in advance of the exercise as possible. This allows maximum time to forecast and stabilize personnel assignments and permits the detachment commanders to make early preparations.

b. Guerrilla Cadre Detachments.

(1) Selection of guerrilla cadre detachments is based upon the experience of the commander and state of training of the detachment. The most experienced detachments are assigned the guerrilla cadre role. The value of the training received by the Special Forces operational detachments depends to a large degree on how thoroughly the guerrilla cadre accomplished its tasks.

(2) Guerrilla cadre detachments are assigned their areas early in the planning stage. This permits the guerrilla cadre to study and reconnoiter the area in advance. If possible, detachments having previous knowledge of the area are assigned to the same area.

c. Operational Detachments.

(1) Operational detachments are selected from those having the least experience.

(2) Personnel assignments are stabilized to allow the operational detachment maximum training as a unit. Personnel of B and C detachments who are engaged in nonoperational duties are released to permit training with their units.

(3) Operational detachments receive area studies of the entire exercise area early in the planning stage. If they do not have physical access to the exercise area, they may be given their area assign-

ments during the preparatory phase. If operational detachments have physical access to the area from their home station, then area assignments are withheld until detachments are isolated at the SFOB. This prevents the operational detachment from having unrealistic advance contact with the exercise.

35. Guerrilla Cadre Detachment Reconnaissance

a. Funds should be provided for support of this reconnaissance and for later operational needs. Guerrilla cadre detachments usually pay for certain services provided by civilians in support of the exercise. These services are essential to the realistic conduct of the exercise. Examples are—

(1) Limited use of civilian transportation (wagons, animals, etc.).
(2) Use of civilian facilities such as barns, sheds, and huts for storage of supplies.
(3) Support of guerrilla cadre detachment personnel who act in the capacity of civilian resistance leaders.
(4) Purchase of food from local sources to supplement issue rations.

b. During the reconnaissance phase, the guerrilla cadre detachment establishes the framework of an auxiliary force. Civilian support is used to the maximum. Various intelligence, warning, and other support mechanisms are prearranged. This is necessary since the exercise is too short to permit the establishment of these support systems during its execution. Thus the guerrilla commander is able to respond to the requirements for civilian support as they develop.

c. The guerrilla commander prepares and submits his reconnaissance report to the control headquarters or field control. These reports are reviewed by the exercise director. The director should discuss the report with each guerrilla commander to ensure the best implementation of exercise requirements. Guerrilla commanders explain their initial physical organization and possible courses of action in the area. Thus, the exercise director is better able to understand the tactical possibilities in the field and to react appropriately.

36. Intelligence

a. Intelligence play is provided by procuring intelligence personnel from outside sources, to be attached to the operational detachments in the briefing center. The individual playing this role should have reconnoitered the area and be qualified to infiltrate with the detachment. Such personnel, if selected from the guerrilla cadre, will have completed the area reconnaissance during the preparatory phase.

b. The amount and kind of information furnished to operational detachments through these intelligence personnel is prescribed by the exercise director. Personnel playing these parts are technically one-time inhabitants of the area and are expected to know the terrain and some leading personalities. Among the latter may be one or more resistance leaders (guerrilla cadre detachment personnel). Using their personal knowledge of the area, they assist in planning the infiltration.

37. Area Organizational Requirements and Restrictions

a. In order to exercise the operational detachments under relatively uniform conditions, the director prescribes certain common area organizational requirements. These requirements, implemented by the guerrilla force, provide each operational detachment with a similar initial situation. In light of this situation, operational detachments then face the problem of obtaining maximum benefit from resistance forces in accomplishing their mission. The situation usually undergoes two or three changes during the course of the exercise.

b. The guerrilla force in each area is divided into two or three widely dispersed units. The number of guerrillas available in each area will determine the number of separate subunits to be established. Thus, operational detachments are faced with the problem of control, logistical support, and training of these separated elements. They also accrue the advantages of additional area coverage, independent operations, and security. Guerrilla commanders may use this arrangement to simulate recruiting or presence of rival bands. Operational detachments are not required to retain this status quo for the duration of the exercise if a change appears necessary.

c. Since guerrilla cadre detachments are authorized to use civilian clothing, the operational detachment should organize and train the resistance elements to accomplish clandestine duties. This restriction will discourage detachment personnel from performing unilateral operations independent of the guerrillas.

d. The use of mechanical transportation by the area command elements may be restricted until physical linkup with field army elements. This restriction forces the guerrillas to depend upon means of transportation which they can realistically expect in operational situations in remote areas. It also teaches personnel to consider time and space in their planning.

e. Another restriction that may be imposed concerns the use of civilian buildings to billet troops. Widespread use of barns, sheds, and other buildings as shelter for participants leads to laxity in security and possible claims for damages against the government. This restriction should not include the use of buildings to shelter sick or convalescent personnel.

38. Operation Plan and Standard Operation Procedures

a. The operation plans for the detachments allow freedom of action, yet contain sufficient detail to provide guidance during periods when communications to the SFOB are sporadic or nonexistent.

b. The operation plan contains as a minimum the following annexes:
- (1) Intelligence.
- (2) Signal.
- (3) Administrative.
- (4) Air support.
- (5) Psychological operations.
- (6) Civil affairs—this annex contains political guidance to assist detachment commanders in dealing with political problems which can be foreseen.
- (7) Evasion and escape.
- (8) Fire support annex.

c. Linkup instructions are delivered to detachments through the field army liaison officer when he joins the guerrillas before linkup. After linkup the field army liaison officer issues the demobilization plan to the detachment.

d. There are two methods used by the SFOB to have the operational detachments initiate operations—
- (1) Execution at the detachment commander's discretion.
- (2) Execution on order from SFOB or based upon a time period or contingent action.

In selecting the method to be used the director considers the probability that communications with the detachments may be interrupted. Initiation of operations should not be entirely dependent upon good communications.

e. Certain theater priority missions are contained in the operation plan. These missions, such as coordinated attacks against lines of communication, and the attack of key installations, are outlined in sufficient detail so that operational detachments may begin planning before deployment. A time limit for execution may be imposed or the action initiated on order of the SFOB. In the latter case, code names are assigned. Transmission of the code name and the time of attack initiates the mission.

f. A standing operating procedure is prepared to govern operational detachment action during the exercise. When possible, this SOP conforms to procedures prescribed for operational use. The SOP includes among other things—

(1) Standard linkup instructions.
(2) Air-support procedures to include DZ and LZ selection and marking.
(3) Logistical procedures.
(4) Demolitions simulation procedures.
(5) Evasion and escape techniques and identification procedures.
(6) Reports, to include format and frequency.

g. The SOP contains minimum report requirements and prescribes simple formats for transmitting messages. A brevity code substituting a word for an entire preselected phrase facilitates reporting. A system for target identification and priority assignment is established to simplify radio communications.

h. Written material required by operational detachments in the field should be miniaturized by photographic means.

i. The SOP is issued well in advance of the preparatory phase to allow operational detachments to become familiar with its contents. The operation plan is issued to detachments in the briefing center.

39. Deployment

Each operational detachment prepares two deployment plans. The primary plan is based upon the assumption that a reception committee is present on the DZ and that no aggressor will intervene. The alternate plan is based upon the assumption that contact with the reception committee is not made because the detachment is dropped in the wrong place or due to aggressor interference on the DZ or some other unforeseen circumstance. It is recommended that alternate plans for one or two detachments be tested. This is accomplished by prearrangement with the guerrilla commander concerned and no contact with guerrillas take place on the DZ. The operational detachment, using its prearranged plan and the intelligence asset, seeks to establish contact. In the event contact is not effected within

36 hours, the guerrillas are administratively directed to the operational detachment.

40. Operations

Certain requirements are levied upon each detachment. These requirements are stated in the operation plan and are designed to allow each detachment to plan and supervise various unconventional warfare missions.

a. At least one large-scale raid is required in each area or sector. Targets are established and guarded by aggressor. Targets may be missile-launching sites, air-defense control installations, nuclear storage or supply depots, and other key installations likely to exist in the operational area. Targets are represented by mockups. Sufficient aggressor guards are located at each installation to require commitment of most of the guerrilla force. A time limit for completion of this attack is established in the operational plan. Accomplishment of the mission no later than the 16th day of the exercise is a suggested time limit. This allows aggressor use of his guard details in other tasks after targets have been attacked and also afford him sufficient time to react effectively to the attack. Also the guerrilla force is available for other tasks subsequent to this operation.

b. A minimum of two ambushes is required of each area. These are coordinated through the guerrilla commander by field control. In conducting an ambush, a guerrilla commander prescribes the route to be traveled by the convoy and a 2-or 3-hour time block within which the convoy is to arrive at the starting point. Aggressor furnishes convoy security. Direction of movement and exact time of arrival of the convoy along the ambush route are not given to the guerrilla commander. This is considered realistic since intelligence of convoy movements is seldom exact. Each convoy is accompanied by a field control representative to assess the effectiveness of the ambush. If the ambush is determined to be unsuccessful, the guerrillas are not allowed to recover any supplies or personnel in the convoy. These ambushes are accomplished in phases four and five.

c. An interdiction mission against lines of communications is established and assigned a code name in the operation plan. This mission is designed to test the effectiveness of area commands and is coordinated by B and C detachments. Detachment planning commences in the briefing center and continues during phase four. The mission is initiated on order of the SFOB. The order to execute is given sufficiently in advance of the execution date to allow the B or C detachment to notify subordinate units. Three or four days is suggested. To permit completion, the operation should take place on two successive dates. This coordinated interdiction operation does not preclude independent LOC attacks earlier in the exercise.

d. Other operational requirements may be included by the director; however, he should be careful not to subdue the initiative of the guerrilla force for independent action.

41. Field Army Control of Guerrilla Forces

a. Operational control of all areas is passed from SFOB to the field army commander on the 26th day of the exercise. SFOB usually plays the role of the field army although field control may also be used.

b. Field army liaison parties join the area command detachments. The liaison party issues instructions for missions in direct support of field army operations. Each area command receives a mission. The liaison party also issues linkup instructions.

c. CBR warfare is integrated into the exercise during the direct support mission. Training chemical agents or smoke may be used against guerrilla forces since they are now in larger groups in relatively fixed positions. Guerrillas are equipped with protective masks procured as a result of an earlier ambush or raid. Simulated radioactive areas can be used to force guerrillas to bypass such areas enroute to their tactical positions.

d. One-half of the aggressor force is employed as friendly conventional troops to affect linkup with the guerrillas. The remainder of the aggressor force is used to test the guerrillas' ability to perform their assigned direct support missions.

e. Upon completion of linkup between the guerrillas and friendly forces, the field army commander orders demobilization of the guerrilla forces. Liaison parties issue specific instructions concerning demobilization procedures. Since this phase is designed to acquaint operational detachments with demobilization problems, the guerrilla commanders create situations such as desertions and hiding arms which require action on the part of the detachment. Phase six terminates with the actual movement of the guerrillas to the base where appropriate ceremonies are conducted. Guerrillas are awarded testimonials of service and certain outstanding personnel receive special recognition. As a part of demobilization, the guerrillas can be administered a short examination to determine the efficiency of training received from the Special Forces operational detachments. Upon completion of the demobilization, guerrilla troops are returned to their parent organizations.

42. Psychological Operations

a. Psychological operations activities are integrated into all phases of the exercise.

b. The guerrilla cadre detachments psychologically prepare the operational areas for the exercise during reconnaissance in the preparatory phase. They attempt to gain support of the civilian population for the friendly forces. They may distribute leaflets to assist them in accomplishing this.

c. Operational detachments monitor psychological operations according to guidance provided in the operation plan. These activities are intended to gain support for the resistance movement and to assist the detachment in accomplishing its mission. The detachment may bring in, or call for, standard leaflets. A proclamation, which recognizes the legal position of the guerrillas, can be introduced in the name of the theater commander. Small portable printing equipment is used for field production of leaflets (see appendix II). A piece of gelatin roll can be used to print newspapers and leaflets. News broadcasts are monitored by the detachment radio or by civilian radio receiver. The objective of the friendly psychological campaign is immediate support of ,area operations and not necessarily broader theater objectives.

d. Aggressor employs a psychological operations unit to support his counter-guerrilla operations. He attempts to separate the guerrillas from their civilian sympathizers and to cause disaffection among guerrilla units. Aggressors may use leaflets, loudspeakers, and other psychological media apropriate to the situation.

e. The exercise director establishes guide lines concerning the use of leaflets and loudspeakers. Before release, all written material should be reviewed by trained psychological operations officers or a representative of the control headquarters.

43. Logistics

a. The exercise director takes the necessary action to insure that logistical support operations are realistic. Items to support operations in the field are provided by the sponsor through the detachments. Actual delivery of this material is accomplished by using the support agencies available.

b. A figure of 60 percent successful missions in air operations is excellent. Air missions are flown approximating actual conditions insofar as flying safety regulations permit. Special waivers are often required to provide air support essential to unconventional warfare operations.

c. The ambush and raid are excellent substitute means for obtaining supplies in case of continued inclement weather. The director controls the use of substitute means to avoid over-supplying the guerrillas and to curb the tendency on the part of operational detachments to bypass the air support plan.

44. Damage Simulation

a. A damage simulation system is established to provide a means of testing destructive techniques. The system substitutes rope for detonating cord, ration boxes or sandbags for actual demolitions, and cardboard tags showing firing systems and weights of charges. In some areas a small live charge can be set off to simulate detonation (0.225 kilograms is recommended).

This is desirable to alert aggressor and give him a chance to react.

b. Targets have the simulated charges and firing systems placed on them as realistically as possible. In some cases this is not feasible since the simulated charges would interfere with traffic (mines, cratering charges, and pressure charges on bridges).

c. Advance notice of targets to be attacked and time of attack is given by guerrilla commanders to field control. Field control directs TDAG to dispatch a team to assess damage. TDAG arrives at the target not sooner than 15 minutes after the time of attack, so as not to compromise the guerrillas. TDAG submits to field control a written report of the estimated effectiveness of the attack. These reports are used to support critiques.

d. Field control notifies aggressor of the attack shortly after it has taken place. This notification serves to give aggressor information he logically would learn in an area he occupies.

e. The damage simulation system, properly applied, is a valuable means for gauging the effectiveness of demolitions training.

45. Evasion and Escape

a. Evasion and escape operations are integrated into the exercise. Ten to fourteen days are allotted for evasion and escape play. In preparing the evasion and escape plan, the director ensures that it does not dominate the exercise to the detriment of other training aspects. Evasion and escape is conducted incidental to the guerrilla warfare mission.

b. The guerrilla commander selects rendezvous areas to contact such personnel, based on his preliminary phase reconnaissance.

c. Evaders are transported into the areas and dropped off at random. The evader should have the basic survival equipment which normally would be available to him. Evaders then attempt to establish contact with the guerrillas.

d. The guerrillas may use the services of local civilians to help locate evaders. Elaborate evasion organizations employing guerrillas are avoided.

e. After recovery, the evaders are identified by the detachment and evacuated by air or boat. If such evacuation fails, evaders may be evacuated administratively.

46. Communications

a. The communications plan provides for operational and administrative communications systems.

b. The operational communications systems uses the signal company personnel and equipment. The operational signal plan conforms as closely as possible to planned wartime procedure. SFOB employs blind transmission broadcasts to the field. Call signs, frequencies, and transmission times are changed with each contact. Detachment mandatory contacts are minimized, one per day is suggested. Contact from the field is accomplished either by scheduled contact or use of a guard frequency. Average transmission times from the field are held to a minimum, consistent with operational necessity.

c. Confirmation of air resupply drops and other prearranged activity is transmitted by blind transmission broadcast using detachment communications.

d. The administrative communications system is used to connect the exercise director with field control and the SFOB; in turn, field control communicates administratively with aggressor, TDAG, ALG, and the guerrilla commanders.

e. The guerrilla commanders contact field control daily on a prearranged schedule. Telephone is the best system for this contact since it does not require extra personnel and usually offers the calling party maximum latitude in selecting the calling point. Commercial telephone service is most commonly used since the facilities are in place. Calls are sent collect to field control.

APPENDIX V
AREA STUDY

Section I. INTRODUCTION

1. General

This appendix is an outline for the preparation of an area study. This guide should be used for research and study of a selected country. The outline provides a systematic consideration of the principal factors which influence Special Forces operational planning.

2. Purpose

The purpose of the area study guide is to provide a means for acquiring and retaining essential information to support operations. Although the basic outline is general in nature, it provides detailed coverage of a given area. As time is made available for further study, various subjects should be subdivided and given to detachment members to produce a more detailed analysis of the area.

3. Techniques of Preparation

Maximum use should be made of graphics and overlays. Most subsections lend themselves to production in graphical or overlay form.

Section II. GENERAL AREA STUDY

4. General

a. Political.
   (1) Government, international political orientation, and degree of popular support.
   (2) Attitudes and probable behavior of identifiable segments of the population toward the United States, its allies, and the enemy.
   (3) National historical background.
   (4) Foreign dependence or alliances.
   (5) National capital and significant political, military, and economic concentrations.

b. Geographic Positions.
   (1) Areas and dimensions.
   (2) Latitude and climate.
   (3) Generalized physiography.
   (4) Generalized land use.
   (5) Strategic location.
      (*a*) Neighboring countries and boundaries.
      (*b*) Natural defenses including frontiers.
      (*c*) Points of entry and strategic routes.

c. Population.
   (1) Total and density.
   (2) Breakdown into significant ethnic and religious groups.
   (3) Division between urban, rural, or nomadic groups.

  (*a*) Large cities and population centers.
  (*b*) Rural settlement patterns.
  (*c*) Areas and movement patterns of nomads.

d. National Economy.
 (1) Technological standards.
 (2) Natural resources and degree of self-sufficiency.
 (3) Financial structure and dependence upon foreign aid.
 (4) Agriculture and domestic food supply.
 (5) Industry and level of production.
 (6) Manufacture and demand for consumer goods.
 (7) Foreign and domestic trade and facilities.
 (8) Fuels and power.
 (9) Telecommunications and radio systems.
 (10) Transportation adequacy by U.S. standards.
  (*a*) Railroads.
  (*b*) Highways.
  (*c*) Waterways.
  (*d*) Commercial air installations.

e. National Security.
 (1) Center of political power and the organization for national defense.
 (2) Military forces (army, navy and air force); summary of order of battle.
 (3) Internal security forces and police

forces; summary of organization and strength.

(4) Paramilitary forces; summary of organization and strength.

5. Geography

a. Climate. General classification of the country as a whole with normal temperatures, rainfall, etc., and average season variations.

b. Terrain. General classification of the country noting outstanding features, i.e., coasts, plains, deserts, mountains, hills, plateaus, rivers, and lakes.

c. Major Geographic Subdivisions. Divide the country into its various definable subdivisions, each with generally predominant topographical characteristics, i.e., coastal plains, mountainous plateau, rolling, and heavily forested hills. For each subdivision use the following outline in a more specific analysis of the basic geography:

(1) *Temperature.* Variations from normal and, noting the months in which they may occur, any extremes that would affect operations.

(2) *Rainfall and snow.* Same as (1) above.

(3) *Wind and visibility.* Same as (1) above.

(4) *Relief.*

(*a*) General direction of mountain ranges or ridge lines and whether hills and ridges are dissected.

(*b*) General degree of slope.

(c) Characteristics of valleys and plains.
  (d) Natural routes for, and natural obstacles to, cross-country movement.

(5) *Land use.* Note any peculiarities, especially—
  (a) Former heavily forested areas subjected to widespread cutting or dissected by paths and roads; also the reverse; pasture or waste land which has been reforested.
  (b) Former waste or pasture land that has been resettled and cultivated and is now being farmed; also the reverse; former rural countryside that has been depopulated and allowed to return to waste land.
  (c) Former swamp or marsh land that has been drained; former desert or waste land now irrigated and cultivated; and lakes created by post-1945 dams.
  (d) Whenever not coincidental with (a), (b), or (c) above, any significant change in rural population density since 1945 is noted.

(6) *Drainage.* General pattern.
  (a) Main rivers, direction of flow.
  (b) Characteristics of rivers and streams such as width, current, banks, depths, kinds of bottoms and obstacles, etc.
  (c) Seasonal variation, such as dry beds and flash floods.
  (d) Large lakes or areas of many ponds

or swamps (potential LZ's for amphibious aircraft).
- (7) *Coast.* Examine primarily for infiltration, exfiltration, and resupply points.
  - (a) Tides and waves, to include winds and current.
  - (b) Beach footing and covered exit routes.
  - (c) Quiet coves and shallow inlets or estuaries.
- (8) *Geological basics.* Types of soil and rock formations (include areas for potential LZ's for light aircraft).
- (9) *Forests and other vegetation.*
  - (a) Natural or cultivated.
  - (b) Types, characterstics, and signficant variations from the norm at the different elevations.
  - (c) Cover or concealment to include density; seasonal variation.
- (10) *Water.* Ground, surface, seasonal, potability.
- (11) *Subsistence.*
  - (a) Seasonal or year-round.
  - (b) Cultivated (vegetables, grains, fruits, and nuts).
  - (c) Natural (berries, fruits, nuts, and herbs).
  - (d) Wild life (animals, fish, and fowl).

6. People

The following outline should be used for an analysis of the population in any given region or country, or as the basis for an examination of the people within a subdivision as suggested in paragraph 5c. Particular attention should be given to those areas within a country where the way of life and the characteristics of the local inhabitants are at variance in one or more ways from the more prevalent, national way of life.

a. Basic Racial Stock and Physical Characteristics.

    (1) Types, features, dress, and habits.

    (2) Significant variations from the norm.

b. Standard of Living and Cultural (Education) Levels.

    (1) Primarily note the extremes away from average.

    (2) Class structures (degree of established social stratification and percentage of population in each class).

c. Health and Medical Standards.

    (1) Common diseases.

    (2) Standards of public health.

    (3) Medical facilities and personnel.

    (4) Potable water supply.

    (5) Sufficency of medical supplies and equipment.

d. Ethnic Components. This should be analyzed only if of sufficient size, strength, and established bonds to constitute a dissident minority of some consequence.

- (1) Location or concentration.
- (2) Basis for discontent and motivation for change.
- (3) Opposition to majority or the political regime.
- (4) Any external or foreign ties of significance.

e. Religion.
- (1) Note wherein the national religion definitely shapes the actions and attitudes of the individual.
- (2) Religious divisions. Major and minor religious groups of consequence. See *d*(1) through (4) above.

f. Traditions and Customs (particularly taboos). Note wherever they are sufficiently strong and established that they may influence an individual's actions or attitude even during a war situation.

g. Rural Countryside.
- (1) Peculiar or different customs, dress, and habits.
- (2) Village and farm buildings; common construction materials.

h. Political Parties or Factions.
- (1) If formed around individual leaders or based on established organizations.
- (2) If a single dominant party exists, is it nationalistic in origin or does it have foreign ties?

(3) Major legal parties with their policies and goals.
(4) Illegal or underground parties and their motivation.
(5) Violent opposition factions within major political organizations.

i. Dissidence. General active or passive potential, noting if dissidence is localized or related to external movements.

j. Resistance (identified movements). Areas and nature of activities, strength, motivation, leadership, reliability, possible contacts, and external direction or support.

k. Guerrilla Groups. Areas and nature of operations, strength, equipment, leader's reliability, contacts, and external direction or support.

7. Enemy

a. Political
  (1) *Outside Power.* Number and status of nonnational personnel, their influence, organization, and mechanism of control.
  (2) *Dominant National Party.* Dependence upon and ties with an outside power; strength, organization, and apparatus; evidences of dissension at any level in the party; and the location of those areas within the country that are under an especially strong or weak nonnational control.

b. Conventional Military Forces (army, navy, and air force).

(1) Nonnational or occupying forces in the country.
   (a) Morale, discipline, and political reliability.
   (b) Personnel strength.
   (c) Organization and basic deployment.
   (d) Uniforms and unit designations.
   (e) Ordinary and special insignia.
   (f) Leadership (officer corps).
   (g) Training and doctrine.
   (h) Equipment and facilities.
   (i) Logistics.
   (j) Effectiveness (any unusual capabilities or weaknesses).
(2) National (indigenous) forces (army, navy, air force). See (1)(a) through (j) above.

c. Internal Security Forces (including border guards).
   (1) Strength and general organization, distinguishing between nonnational and national elements.
      (a) Overall control mechanism.
      (b) Special units and distinguishing insignia.
      (c) Morale, discipline, and relative loyalty of native personnel to the occupying or national regime.
      (d) Nonnational surveillance and control over indigenous security forces.
      (e) Vulnerabilities in the internal security system.
      (f) Psychological vulnerabilities.

- (2) Deployment and disposition of security elements.
    - (a) Exact location down to the smallest unit or post.
    - (b) Chain of command and communication.
    - (c) Equipment, transportation, and degree of mobility.
    - (d) Tactics (seasonal and terrain variations).
    - (e) Methods of patrol, supply, and reinforcements.
- (3) The location of all known guardposts or expected wartime security coverage for all types of installations, particularly along main lines of communication (LOC) (railroads, highways, and telecommunication lines) and along electrical power and POL lines.
- (4) Exact location and description of the physical arrangement and particularly of the security arrangements of all forced labor or concentration camps and any potential PW inclosures.
- (5) All possible details, preferably by localities, of the types and effectiveness of internal security controls, including checkpoints, identification cards, passports, and travel permits.

8. Targets

The objective in target selection is to inflict maximum damage on the enemy with minimum expenditure of men and material, and without counterproductive or significantly adverse effect upon the friendly elements of the population. Initially, the operational capabilities of a guerrilla force may be limited in the interdiction or destruction of enemy targets. The target area and the specific points of attack must be studied, carefully planned, and priorities established. In general, targets are listed in order of priority.

   a. Railroads.
   - (1) Considerations in the selection of a particular line.
      - (*a*) Importance, both locally and generally.
      - (*b*) Bypass possibilities.
      - (*c*) Number of tracks and electrification.
   - (2) Location of maintenance crews, reserve repair rails, and equipment.
   - (3) Type of signal and switch equipment.
   - (4) Vulnerable points.
      - (*a*) Unguarded small bridges or culverts.
      - (*b*) Cuts, fills, overhanging cliffs, or undercutting streams.
      - (*c*) Key junctions or switching points.
      - (*d*) Tunnels.
   - (5) Security system.

   b. Telecommunications.

   c. POL.

   d. Electric Power.

   e. Military Storage and Supply

   f. Military Headquarters and Installations.

g. Radar and Electronic Devices.

  h. Highways.

  i. Inland Waterways—Canals.

  j. Seaports.

  k. Natural and Synthetic Gas Lines.

  l. Industrial Plants.

  Note. Targets listed in *b* through *l* above are divided into subsections generally as shown in *a* above. Differences in subsections are based upon peculiarities of the particular target system.

Section III. INTELLIGENCE STUDY GUIDE

9. General

The following guide is designed to bring the essentials of operational area intelligence into focus. It is built upon section II, this appendix, but narrows the factors to apply to a relatively small and specific area. The guide refines the critical elements and places them into proper perspective of an actual operation at a given time.

10. Considerations

a. Select from section II those elements that are applicable to the situation and the assigned guerrilla warfare operational area. Use appropriate parts of paragraphs 5–8.

b. Eliminate nonessential data; prepare a brief, concise summation of basic facts.

c. Note serious gaps in data as processed in 10*b* above and take immediate action to obtain the most current reliable information.

d. Prepare or request graphics; large-scale sheets and special maps covering the assigned area; the latest photograph and illustration or information sheets on targets within the area; town plans, sketches of installations, and air and hydrographic charts related to the area.

e. Assemble the material for ready reference. Plot on maps or over-lays the following information as developed:

(1) Recommended initial guerrilla bases and alternate bases.

(2) Primary and alternate DZ's, LZ's, or points for other forms of infiltration.

(3) Possible direction and orientation points for infiltration vehicles (aircraft or boat), landmarks, etc.

(4) Routes from infiltration point to likely guerrilla base with stopover sites.

(5) Points for arranged or anticipated contacts with friendly elements.

(6) Enemy forces known or anticipated to include location, strength, and capabilities.

(7) Estimate of enemy operations or movements during the infiltration period.

(8) Settlements or scattered farms in the vicinity of the infiltration point and tentative guerrilla bases.

(9) All railroads, highways, telecommunications, etc., in the GWOA.

(10) All important installations and facilities.

(11) Significant terrain features.

(12) Off-road routes and conditions for movement in all directions.
   (13) Distances between key points.
   (14) Recommended point of attack on assigned target system, and selection of other potential target areas.

f. Continue to collect information and revise estimates in keeping with more current intelligence. Develop increasing detail on *e*(1) through (14), above.

g. Emphasize obtaining information on the following:

   (1) Local inhabitants.
      (*a*) Ethnic origins and religion.
      (*b*) Local traditions, customs, and dress.
      (*c*) Food, rationing, and currency.
      (*d*) Attitudes toward the regime, the United States, and for or against existing political ideologies.
      (*e*) Any peculiarities or variances among individuals or small groups.
      (*f*) Group leadership and systems of control or influence employed.
   (2) Enemy military forces and installations.
   (3) Internal security forces and police.
      (*a*) Organizations, locations, and strengths.
      (*b*) Unit designations, insignia, and uniforms.
      (*c*) Areas covered and unit responsibilities.

(d) Checkpoints, controls, and current documentation.
   (e) Patrols and mobile units.
(4) Geographic features in maximum detail.
(5) Approaching seasonal climatic changes and their effect upon weather and terrain.
(6) Target categories and target areas in greater detail.

APPENDIX VI
AREA ASSESSMENT

Section I. GENERAL

1. General.

a. In order to plan and direct operations, Special Forces detachment commanders need certain basic information about the operational area. This information, when gathered or confirmed in the operational area, is called an area assessment.

b. An area assessment is the immediate and continuing collection of information started after infiltration into a guerrilla warfare operational area. Characteristically it—

   (1) Confirms, corrects, or refutes previous intelligence of the area acquired as a result of area studies and other sources prior to infiltration.
   (2) Is a continuing process.
   (3) Forms the basis for operational and logistical planning for the area.
   (4) Includes information of the enemy, weather, and terrain.
   (5) Includes information on the differently motivated segments of the civil population in the area of operations.

c. The results of the area assessment should be transmitted to the SFOB only when there is marked deviation from previous intelligence. The SFOB prescribes in appropriate SOP's and annexes those items to be reported.

d. The following outline, containing the major items of interest to the area command, is an example of how such an assessment may be accomplished. This outline is designed to facilitate the collection, processing, and collation of the required material. There are two types of area assessment based upon degrees of urgency—initial and principal.

Section II. INITIAL AND PRINCIPAL ASSESSMENTS

2. Initial Area Assessment

Initial assessment includes those items deemed essential to the operational detachment immediately following infiltration. These requirements must be satisfied as soon as possible after the detachment arrives in the operational area, and should include—

a. Location and orientation.

b. Detachment physical condition.

c. *Overall security.*
  (1) Immediate area.
  (2) Attitude of the local population.
  (3) Local enemy situation.

d. Status of the local resistance elements.

3. Principal Area Assessment

Principal assessment, a continuous operation, includes those collection efforts which support the continued planning and conduct of operations. It forms the basis for all of the detachment's subsequent activities in the operational area.

a. The Enemy.
  (1) Disposition.
  (2) Composition, identification, and strength.
  (3) Organization, armament, and equipment.
  (4) Degree of training, morale, and combat effectiveness.
  (5) Operations.
     (*a*) Recent and current activities of the unit.
     (*b*) Counterguerrilla activities and capabilities with particular attention to reconnaissance units, special troops (airborne, mountain, ranger), rotary wing or vertical lift aviation units, counterintelligence units, and units having a mass CBR delivery capability.
  (6) Unit areas of responsibility.
  (7) Daily routine of the units.
  (8) Logistical support to include—
     (*a*) Installations and facilities.
     (*b*) Supply routes.
     (*c*) Methods of troop movement.
  (9) Past and current reprisal actions.

 b. Security and Police Units.
  (1) Dependability and reliability to the existing regime or the occupying power.
  (2) Disposition.
  (3) Composition, identification, and strength.
  (4) Organization, armament, and equipment.
  (5) Degree of training, morale, and efficiency.
  (6) Utilization and effectiveness of informers.
  (7) Influence on and relations with the local population.
  (8) Security measures over public utilities and government installations.

 c. Civil Government.
  (1) Controls and restrictions, such as—
   (a) Documentation.
   (b) Rationing.
   (c) Travel and movement restrictions.
   (d) Blackouts and curfews.
  (2) Current value of money, wage scales.
  (3) The extent and effect of the black market.
  (4) Political restrictions.
  (5) Religious restrictions.
  (6) The control and operation of industry, utilities, agricultural, and transportation.

 d. Civilian Population.
  (1) Attitudes toward the existing regime or occupying power.

- (2) Attitudes toward the resistance movement.
- (3) Reaction to United States support of the resistance.
- (4) Reaction to enemy activities within the country and, specifically, that portion which is included in guerrilla warfare operational areas.
- (5) General health and well-being.

e. Potential Targets (for each consider population reaction).
- (1) Railroads.
- (2) Telecommunications.
- (3) POL.
- (4) Electric power.
- (5) Military storage and supply.
- (6) Military headquarters and installations.
- (7) Radar and electronic devices.
- (8) Highways.
- (9) Inland waterways and canals.
- (10) Seaports.
- (11) Natural and synthetic gas lines.
- (12) Industrial plants.
- (13) Key personalities.

f. Weather.
- (1) Precipitation, cloud cover, temperature, visibility, and seasonal changes.
- (2) Wind speed and direction.
- (3) Light data (BMNT, EENT, sunrise, sunset, moonrise, and moonset).

g. Terrain.

    (1) Location of areas suitable for guerrilla bases, units, and other installations.

    (2) Potential landing, drop zones, and other reception sites.

    (3) Routes suitable for—
      (*a*) Guerrillas.
      (*b*) Enemy forces.

    (4) Barriers to movement.

    (5) The seasonal effect of the weather on terrain and visibility.

h. Resistance Movement.

    (1) *Guerrillas.*

      (*a*) Disposition, strength, and composition.

      (*b*) Organization, armament, and equipment.

      (*c*) Status of training, morale, and combat effectiveness.

      (*d*) Operations to date.

      (*e*) Cooperation and coordination between various existing groups.

      (*f*) General attitude towards the United States, the enemy, and various elements of the civilian population.

      (*g*) Motivation of the various groups and their receptivity.

      (*h*) Caliber of senior and subordinate leadership.

      (*i*) Health of the guerrillas.

(2) *Auxiliaries or the underground.*
   (a) Disposition, strength, and degree of organization.
   (b) Morale, general effectiveness, and type of support.
   (c) Motivation and reliability.
   (d) Responsiveness to guerrilla or resistance leaders.
   (e) General attitude towards the United States, the enemy, and various guerrilla groups.

i. *Logistics Capability of the Area.*
   (1) Availibility of food stocks and water to include any restrictions for reasons of health.
   (2) Agriculture capability.
   (3) Type and availability of transportation of all categories.
   (4) Types and location of civilian services available for manufacture and repair of equipment and clothing.
   (5) Supplies locally available to include type and amount.
   (6) Medical facilities to include personnel, medical supplies, and equipment.
   (7) Enemy supply sources accessible to the resistance.

Section III.

PRINCIPAL PREVENTIVE MEDICINE AREA ASSESSMENT

4. Climatology and Topography

a. Cold weather.
   (1) How long is the cold weather season?
   (2) Is the weather cold enough to put emphasis on causes, treatment, and prevention of cold weather injuries?
   (3) What is the temperature range?
   (4) Is it a dry cold or a wet cold?

b. Hot weather.
   (1) How long is the hot weather season?
   (2) What is the temperature range?
   (3) Is the weather hot enough to put emphasis on causes, treatment, and prevention of heat injuries?
   (4) Is acclimatization necessary? How many days?
   (5) Describe humidity.

c. Terrain.
   (1) What types of terrain are present?
   (2) What is the range in elevation?
   (3) Are swamps present?
   (4) Are rivers present?
   (5) Describe terrain over which you operate and that on which your camp is located.
   (6) How does the terrain affect evacuation and medical resupply?

5. Indigenous Personnel

   a. Physical characteristics.

   (1) Give names of tribes or groups you associate with.

   (2) Describe indigenes in terms of height, body build, color, and texture of skin and hair (photographs will help).

   (3) Describe endurance, ability to carry loads, and to perform other physical feats.

   b. Dress.

   (1) Describe clothing, principal ornamentation, and footgear.

   (2) What symbolism is attached to various articles of clothing and jewelry, if any?

   (3) Are amulets worn and what do they symbolize? (Again, photographs will be useful.)

   c. Attitudes.

   (1) What taboos and other psychological attributes are present in the society?

   (2) What are attitudes toward birth, puberty, marriage (monogamy, polygamy, polyandry), old-age, sickness, death, etc.? (Describe any rituals associated with these events.)

   (3) Describe attitudes toward doctors and Western medicine.

   (4) Describe rites and practices by witch doctors during illness. What do these

rites symbolize? Does the practitioner use Western medicines?

(5) Do indigenes respond to events in the same manner you would (fear, happiness, anger, sadness)?

d. Housing.

(1) Describe physical layout of the community (diagrams and notes will help).

(2) Describe construction and materials used. Give layout of typical houses (sleeping areas, cooking areas, etc.).

(3) Describe infestation with ectoparasites and vermin.

(4) Describe general cleanliness.

(5) How many persons inhabit a dwelling?

e. Food.

(1) Describe the ordinary diet.

(2) Is food cultivated for consumption? What foods?

(3) Describe agricultural practices (slash and burn, permanent farms, etc.). Is human waste used for fertilizer? What domestic animals are raised for food?

(4) Does hunting or fishing or gathering wild crops contribute significantly to the diet? What wild vegetables are consumed?

(5) How is food prepared? What foods are cooked, pickled, smoked, or eaten raw? How are foods preserved (smoking, drying, etc.)?

(6) At what age are children weaned? What diet is provided infants after weaning occurs?

(7) Which members of the family are given preference at table?

(8) Are there food taboos? Is milk denied children, infants, pregnant or lactating mothers? Why?

(9) How does the season of your (temperature, rainfall, etc.) influence diet? Are there famines? Does migration in search of food occur?

(10) What foods provided by U.S. personnel do indigenes prefer or reject?

(11) What cash crops are raised?

6. Water Supply, Urban

a. Is water plentiful?

b. Is water treated?

c. What kind of water treatment plants are used (if any)?

7. Water Supply, Rural

a. What are the numbers and types of rural water supplies?

(1) What are the sources (rivers, springs, or wells)?

(2) Is same water used for bathing, washing, and drinking?

(3) How far is water source from village?

(4) Is water plentiful or scarce? Give seasonal relationships to water availability. Is water rationed by indigenes?

b. What treatment is given to water in rural areas?

(1) Is water boiled?

(2) Is water filtered or subjected to other purification process before consumption?

(3) Give attitudes of indigenes toward standard U.S. purification methods.

(4) Is the untreated water safe for bathing?

8. Sewage Disposal (When Applicable)

a. What are the types and locations of sewage treatment plants?

b. Do urban areas have combined separate sewage systems?

c. What are the locations of outfalls and conditions of receiving streams before and after receiving affluent? State whether or not affluent is chlorinated and if so, how much.

d. What types of sewage disposal are used in those areas not connected to a sewer system? Give approximate numbers or each—

(1) Septic tanks.

(2) Cesspools, leaching pits, etc.

(3) Sanitary privies.

e. In the village, what system is used for disposal of human excrement, offal, and dead animals, or humans?

> (1) Is excrement collected for use as fertilizer? How?
>
> (2) What scavengers commonly assist in the process of disposal (animals, birds, etc.)?
>
> (3) What is the relationship of disposal sites to watering sites?
>
> (4) What are the attitudes of indigenes to standard U.S. methods, such as the use of latrines?

9. Epidemiology

What specific diseases in each of the following major categories are present among the guerrillas, their dependents, or their animals?

a. Respiratory disease.

b. Arthropod-borne disease.

c. Intestinal disease.

d. Venereal disease.

e. Miscellaneous diseases, such as tetanus, scabies, rabies and dermatophytosis.

f. Relate the occurrence of more than one similar illness at a given time to a common source or event (if diarrhea was associated with a specific meal, get accurate description of types of food served, method of preparation, etc.).

10. Domestic Animals

   a. What domestic animals are present?

   (1) Are they raised for food, labor, or other purposes?

   (2) Of what importance is the herd or flock economically?

   (3) Is any attempt made to breed animals selectively?

   b. Describe the normal forage.

   (1) Do owners supplement the food supply? What food supplements are given, if any?

   (2) Are animals penned, or allowed to roam?

   (3) Where are the animals housed?

   c. Is any religious symbolism or taboo associated with animals ("sacred cows")? Are animals sacrificed for religious purposes?

   d. Describe in detail the symptoms and signs in any diseased animal. Which species were involved?

   e. If disease is epidemic, describe the epidemic?

   (1) What percentage of the herd or flock demonstrated illness?

   (2) Did all of the illnesses occur more or less spontaneously, or over a period of time?

   (3) Was more than one herd or flock affected at the same time and place?

   (4) Did the epidemic spread to neighboring villages? What was the time interval between?

(5) Were young and old animals similarly affected, or just the young, or just the old? Had such outbreaks occurred before?

(6) Was the outbreak related to the introduction of new animals to the herd or flock or to some other observable event? What event? How long after the event(s) did the outbreak occur?

(7) At what season of the year did the outbreak occur?

(8) What percent of animals died?

(9) Were any human illnesses associated with the outbreak? Describe.

(10) What measures, if any, did the owners take to control the outbreak? What was done with sick or dead animals?

(11) Do herds or flocks reproduce normally, or is early abortion common?

(12) Are animals fat and healthy looking, or in generally poor condition?

(13) What symptoms did sick animals show (icterus, emaciation, diarrhea, nasal discharge, swollen lymph glands, or incoordination)?

(14) Are local veterinarians available for animal treatment and ante- and postmortem inspections of meats? What is their training?

(15) If lay treatment of animals is accomplished, what are common treatments and practices?

(16) What is previous vaccination history, if any?

(17) What diseases are endemic to animals in the area?
  (*a*) Anthrax.
  (*b*) Brucellosis.
  (*c*) Tularemia.
  (*d*) Rabies.
  (*e*) Trichinosis.
  (*f*) Worms.
  (*g*) Bovine tuberculosis.
  (*h*) Q-Fever.
  (*i*) Others.

11. Local Fauna

a. Record species of birds, large and small mammals, reptiles, and arthropods present in the area; if names are unknown, describe.

b. Note relationship of these species, including burrows and nesting sites, to human habitation, food supplies, and watering sites.

c. Note the occurrence of dead or dying animals, especially if the die-off involves large numbers of a given species; note relationship of die-off to occurrence of human disease.

d. Report any methods used by indigenes to defend against ectoparasites. How effective were these methods? How effective are standard U.S. protective measures?

12. Poisonous Plants

Record those species which are known to be

toxic to man through contact with the skin, inhalation of smoke from burning vegetation, or through ingestion.

13. Arthropods of Medical Importance (Include Species and Prevalence When Known)

a. Mosquitoes.

b. Other flies.

c. Fleas.

d. Mites.

e. Ticks.

f. Lice.

g. Spiders and other arachnids.

APPENDIX VII
CATALOG SUPPLY SYSTEM

1. General

a. This appendix provides a guide for the planning and preparation of a catalog supply system for Special Forces operations.

b. The catalog supply system shown after paragraph 6*a* in this appendix is a sample only. Special Forces unit commanders should develop a catalog supply system for their units based upon contingency plans and unit field SOP's.

c. The catalog for an operational detachment should be brief, clear, concise, and preferably printed in miniature to facilitate ease in handling.

2. Responsibilities

a. The preparation of a catalog supply system is normally delegated to the Special Forces group S-4.

b. The preparation of supplies for delivery to a GWOA is normally the responsibility of the logistical support element.

3. Characteristics

a. The catalog supply system uses a brevity

code in which a single item or several associated items are identified by a code word.

b. Comprises packages of associated individual items as well as units comprising several packages. This combination permits the user maximum flexibility in requesting supplies.

c. Is based upon the guerrilla organization described in FM 101–10–3.

4. Principles

a. Principles in preparing a catalog supply system are to determine broad classification of supplies, i.e., quartermaster, and signal.

b. Assign code designations to each of these broad classifications.

c. Insure that ancillary supplies are included with the main item, e.g., ammunition and cleaning supplies are included with all weapons; batteries are included with flashlights.

d. Determine packaging components based upon paragraph 1*b* above.

5. Preparation of Supplies

a. Each man-portable package should be equipped with carrying straps to facilitate rapid clearance of reception site.

b. Each package should be waterproof to permit caching above ground, and limited caching underground or underwater.

c. Maximum use should be made of reusable items for packaging materials, such as clothing and blankets for padding and ponchos for waterproofing.

d. Include inventory list with all packages.

e. Insure that any required instructional material is printed in the language of the guerrillas as well as in English.

f. Include selected morale and comfort items, if possible.

g. Insure each package is marked in accordance with a prearranged coding system, so that contents can be readily identified without opening package.

h. The number of delivery containers that can be used is determined by the available means of transportation (aircraft, submarine, etc.).

6. Request Procedure

a. The coding system is not secure by itself, but will reduce message length when a variety of supplies are ordered. For this example, each type of supply is a sequence of assigned letter designations

Section	
I Chemical	ALPHA ALPHA through DELTA ZULU
II Demolition/ Mines	ECHO ALPHA through HOTEL ZULU
III Medical	INDIA ALPHA through LIMA ZULU

Section		
IV	Weapons/Ammunition	MIKE ALPHA through PAPA ZULU
V	Quartermaster	QUEBEC ALPHA through TANGO ZULU
VI	Signal	UNIFORM ALFA through WHISKEY ZULU
VII	Special	XRAY ALPHA through ZULU ZULU

b. To reduce unreadable garble when ordering supplies, maximum use should be made of phonetic spelling. Some units and packages may be followed by a numbered list showing the contents of the package or unit. For these items, the unit or package can be ordered complete, or any numbered item may be ordered separately. For example, clothing and equipment for 40 men is required. Determine the boot sizes needed and include in the message. Assume that the following boot sizes are desired: Ten pair size 8–1/2W, six pair size 9M, three pair size 9–1/2N, four pair size 10N, six pair size 10M, two pair size 10W, five pair size 10–1/2M, four pair size 11M. The message reads—

ONE QUEBEC ALPHA PD BOOTS TEN SIZE EIGHT PT FIVE WHISKEY SIX SIZE NINE MIKE THREE SIZE NINE PT FIVE NOVEMBER FOUR SIZE TEN NOVEMBER SIX SIZE TEN MIKE TWO SIZE TEN WHISKEY FIVE SIZE TEN PT FIVE MIKE FOUR SIZE ELEVEN MIKE.

Clothing should be packed to approximately match boot sizes (section V). On the other hand if only 40 ponchos are desired, the request would read, TWO ZERO QUEBEC ALPHA SEVEN. (Note that item 7 under QUEBEC ALPHA reads *two* ponchos; therefore, 20 each of item 7 would be ordered.)

c. The system should permit all items listed in each unit to be ordered separately if necessary. Normally the complete unit should be ordered.

d. Items not listed should be requested by nomenclature in sufficient detail to insure thorough understanding at the SFOB; for example: TWO GASOLINE LANTERNS.

e. Using the classes of supply as listed in paragraph 6*a*, above, following is a sample format of a catalog supply system. All entries are representative only.

SAMPLE FORMAT
CATALOG SUPPLY SYSTEM

Section I. CHEMICAL

Code	Unit Designation	Unit Weight	No. Pkgs.	Contents
ALPHA ALPHA	Chemical Grenade No. 1 (16 rds)	46 lbs	1	Sixteen grenades, hand, smoke WP, M15 packed in individual containers.
ALPHA BRAVO	Chemical Grenade No. 2 (16 rds)	47 lbs	1	Sixteen grenades, hand, incendiary, (TH) AN, M14 packed in individual containers.
ALPHA CHARLIE	Chemical Grenade No. 3 (16 rds)	34 lbs	1	Sixteen grenades, smoke, colored. M18 (Green, red, violet, and yellow) packed in individual containers.

Section II. DEMOLITIONS AND MINES

Code	Unit Designation	Unit Weight	No. Pkgs.	Contents
ECHO ALPHA	Demolitions No. 1 (20 blocks)	50 lbs	1	20 blocks, demolition, M5A1 (2½ lb comp C-4)
ECHO BRAVO	Demolitions No. 2 (2 assemblies)	44 lbs	1	Two assemblies, demolition M37 (2½ lb comp (C-4) 8 blocks per assembly.
ECHO CHARLIE	Demolitions No. 3	45 lbs	1	45 blocks, demolition, (1 lb TNT)

Section III. MEDICAL

Code	Unit Designation	Unit Weight	No. Pkgs.	Contents
INDIA ECHO	Dental Unit	21 lbs	1	Three dental kits, emergency field (645-927-8440).
INDIA OSCAR	Shock Set No. 2	24 lbs	1	12 bottles of Dextran 500 cc Bottle with injection assembly.

Code	Unit Designation	Unit Weight	No. Pkgs.	Contents
JULIET ALPHA	Orthopedic Cast No. 2	50 lbs	1	1. Three boxes, bandages, cotton, plaster impregnated, 3 in., 12 per box. 2. Three boxes bandages, cotton, plaster impregnated, 4 in., 2 per box. 3. Three boxes, bandages, cotton, plaster impregnated, 6 in., 12 per box.

Section IV. WEAPONS AND AMMUNITION

Code	Unit Designation	Unit Weight	No. Pkgs.	Contents
				Unit Data
MIKE BRAVO	Carbine (20)	240 lbs	4	1. Five carbines, cal. 30, M-2 (30 lbs). 2. Fifteen magazines, carbine, 30-rd capacity (4 lbs). 3. 800 rds cartridge, ball, carbine cal. .30, M-1, packed in ammunition can M-6 (1 can, 25 lbs). 4. One spare parts and accessory packet (2 lbs) (Notes 1 & 2).

See notes at end of table.

Code	Unit Designation	Unit Weight	No. Pkgs.	Unit Data — Contents
MIKE ECHO	Pistol (12)	90 lbs	2	1. Six pistols, automatic, cal. .45, M1911A1 (15 lbs). 2. Eighteen magazines, pistol, cal. .45 (5 lbs). 3. 800 rds, cartridge, ball, cal. .45 packed in ammunition box M5 (1 box, 29 lbs). 4. Six shoulder stocks, pistol (6 lbs).
MIKE HOTEL	Sniper Rifle (6)	165	3	1. Two rifles, cal. .30 M1C, complete (23 lbs). 2. 480 rds, cartridge, AP, cal. 30, 8-rd clips in bandoleers, packed in ammunition can M–8, (2 cans, 32 lbs).

NOTES.

1. The spare parts and accessory packet includes items most subject to damage or wear and tools required for the care and maintenance of the weapons.
2. Weapons units contain cleaning and preserving material, such as rod, lubricants, and patches.

Section V. QUARTERMASTER

Code	Unit Designation	Unit Weight	No. Pkgs.	Contents
QUEBEC ALPHA	Clothing and Equipment—40 personnel. *Notes 1 & 2*	840 lbs	20	Two man unit consisting of: 1. Two belts, pistol OD. 2. Two blankets, OD. 3. Two pair boots, combat. 4. Two coats, man's water resistant sateen (field jacket). 5. Two canteen, dismounted w/cup and cover. 6. Two caps, field, poplin. 7. Two ponchos, coated nylon, OD-107. 8. Two pouches and packets, first aid. 9. Two pair socks, wool. 10. Two pair suspenders, trousers, OD-107. 11. Two pair trousers, men's, cotton, water resistant sateen (field trousers) (42 lbs).

See notes at end of table.

Code	Unit Designation	Unit Weight	Unit Data No. Pkgs.	Controls
ROMEO ALPHA	Rations, Indigenous Personnel—100 men. *Note 3*.	1,750 lbs	35	High fat content meat or canned fish/poultry, sugar, tobacco, salt, coffee or tea, grain flour or rice, accessory items and water purification tablets (50 lbs).
ROMEO ECHO	Packet, barter *Note 4*.	500 lbs	10	50 lb packages.

NOTES.

1. Items vary with the climatic zone and season. This package is based on the temperate zone for spring, summer, and fall seasons. For winter, add gloves and 1 extra blanket per individual.

2. Clothing sizes are issued as small, medium, and large. Clothing is matched to size of boots. Boot size is included in the message requesting the clothing package. The packaging agency dictates matching of boot and clothing sizes based upon experience factors applicable to the operational area.

3. Special rations for indigenous personnel are determined by the area of operations. Allotment is 15 lbs per individual per month.

4. Contents are to be determined by the area of operations.

Section VI. SIGNAL

Code	Unit Designation	Unit Weight	No. Pkgs.	Controls
UNIFORM ALPHA	Batteries No. 1	48 lbs	1	6 BA 279/U for AN/PRC-10.
UNIFORM GOLF	Radio Set AN/PRC-10(1)	42 lbs	1	1. One AN/PRC-10 complete (18 lbs). 2. Three batteries BA 279/U (24 lbs).
UNIFORM MIKE	Radio Set AN/GRC-109(1)	92 lbs	1	1. Radio transmitter, T-784 (9 lbs). 2. Radio receiver, RR-1004 (10 lbs). 3. Power supply, PP-2684 (25 lbs). 4. Operating spares and accessories (6 lbs). 5. Generator, G-43/G, complete (22 lbs). 6. Adapter, CN-690 (2 lbs). 7. 16 batteries, BA 317/U (16 lbs).

Section VII. SPECIAL

Code	Unit Designation	Unit Weight	No. Pkgs.	Unit Data — Controls
XRAY ALPHA	River Crossing Unit No. 1	50 lbs	1	1. Five life rafts, inflatable, one-person capacity with CO_2 cylinder and accessory kit. 2. Five life preservers, yoke, with gas cylinder. 3. Five paddles, boat, 5 feet long.
XRAY BRAVO	River Crossing Unit No. 2	50 lbs	2	1. One life raft, inflatable, seven-person capacity, with CO_2 cylinder and accessory kit. 2. Seven life preservers, yoke with gas cylinders. 3. Four paddles, boat, 5 feet long.

APPENDIX VIII

EXAMPLES OF MASTER TRAINING PROGRAM FOR INDIGENOUS FORCES

1. Example of a master training program for a 10-day leadership school.

Subject	Scope	Hours Day	Hours Night	Hours Total	PE
Map Reading and Compass.	Same general scope as in 30-day program. Include how to read map scale and coordinates.	4	2	6	(4)
First Aid, Field Sanitation, and Survival.	Same general scope as in 30-day program. Emphasis on field sanitation and responsibility of commanders.	4		4	(1)
Individual Tactical Training (Day and Night).	Same general scope as in 30-day program. Emphasis on security of operational bases, movements, formations, control measures at night, and duties and responsibilities of commanders.	10	9	19	(16)
Patrols, Small Unit Tactics, Raids,	Same general scope as in 30-day program. Emphasis on planning,	10	20	30	(25)

Subject	Scope	Day	Night	Total	PE
Ambushes (Day and Night).	organization, preparation, command, control, security, and execution of patrols, ambushes and raids.				
Weapons (U.S. and Foreign).	Same general scope as in 30-day program. Familiarization firing. Primary emphasis on employment of weapons.	8	2	10	(7)
Intelligence	Same general scope as in 30-day program. Primary emphasis on intelligence gathering methods, systems, and counterintelligence. Night vision.	6	4	10	(8)
Air Operations	Same general scope as in 30-day program. Primary emphasis on selection and reporting of DZ's, organization of reception committee, duties and responsibilities of commanders.	6	8	14	(11)
Demolition	Familiarization with demolition	5	-----	5	(3)

	procedures. Demonstration, planning, safety.			
Communications	Communication means, available systems, communication security, simple cryptographic systems.	4	---	4	(2)
Leadership Principles and Techniques.	Military leadership, traits, principles, indications, actions, and orders. Responsibilities and duties of the commander. Human behavior problem areas and problem solving process. Selection of junior leaders. Span of control and chain of command. Combat leadership.	6	---	6	(4)
Tactics and Operations.	Characteristics of guerrilla warfare, guerrilla operations, principles, capabilities, and limitations, organization of operational bases, security, civilian support, logistics, counterintelligence, combat employment, missions, tactical control measures, target selection, mission support site, and defensive meas-	7	7	12	(9)

Subject	Scope	Hours			
		Day	Night	Total	PE
	ures. Responsibilities and duties of indigenous leaders.				
	Total hours within master training program.	70	50	120	(90)

NOTES:

a. Identify those personnel whose leadership ability, knowledge, skill, or desire is below acceptable standards.

b. One additional day may be scheduled upon completion of leadership school for coordinating and planning future operations.

c. The 10-day master training program for the leadership school was developed to provide the indigenous leaders and potential leaders with a general knowledge of the subjects to be taught to all indigenous personnel. The primary emphasis was placed on the role of the leader or commander in order to prepare these leaders to supervise the activities of their subordinates. In this example it is assumed that most of these personnel have had prior military service and therefore should already possess a basic knowledge of the subjects to be covered. Upon completion of the 10-day leadership school, the leaders will return to their units of work and train with their units, thus expanding their own knowledge of the subjects covered.

d. A suggested arrangement of scheduling is as follows:
29 April–4 May: Preparation for training and selection of leaders.
5–14 May: Leadership training.
16 May–14 June: Troop training.

2. An example of a 30-day master training program which may be used as a basis for preparing individual master training programs for each separate indigenous unit.

Subject	Scope	Day	Night	Total	PE
Map Familiarization and Use of Compass.	How to read a map, orientation of map with compass, how to locate oneself, determine azimuth, and day and night use of compass.	14	10	24	(20)
First Aid, Field	Basic treatment of wounds, prevention of infection, simple bandaging, pressure points, prevention of shock, splints, litter construction, and use. Field sanitation measures regarding water supply, waste disposal, and personal hygiene.	6	4	10	(7)
Individual Tactical Training (Day and Night).	Camouflage, cover, concealment, movement, observation, reporting, discipline, sounds, hand-to-hand combat, combat formations, night movement, night camou-	26	9	45	(41)

Subject	Scope	Hours			
		Day	Night	Total	PE
	flage, preparation of equipment and clothing, night vision, sounds and observation, night security and formations, message writing, immediate action drills, and security of operational bases.				
Patrols, Small Unit Tactics, Raids, Ambushes (Day and Night).	Planning, organization, preparation, formations, commands, control, security, communications and reporting of patrols; objectives, target selection, organization of raid forces; reconnaissance and intelligence; planning, preparation, movement, deployment, conduct of raids, disengagement and withdrawal of raiding forces; characteristics, definition, objectives of ambushes, selection of ambush sites, organization of ambush forces,	26	44	70	(60)

	phases of ambush operation, planning, preparation movement, deployment, execution, disengagement, and withdrawal of ambush forces. All subjects covered for both day and night operations.			
Weapons (Foreign and U.S.).	Carbine, M1; SMG; AR; Pistol, .45; machine guns; foreign weapons to include care and cleaning, loading, aiming, stoppages, range firing; familiarization firing of all weapons; and day and night firing.	28	10	38	(32)
Intelligence	Security measures, how to obtain and report information, sources of information, captured documents and material, interrogation and handling of prisoners, and counterintelligence procedures.	8	---	8	(5)
Air Operations	Establishment of drop zone, marking and identification of DZ, security of DZ, receiving and	16	15	31	(25)

Subject	Scope	Hours			
		Day	Night	Total	PE
	transporting supplies and equipment.				
Demolitions	Non-electric and electric firing systems, calculation and placement of charges, rail and bridge destruction, booby traps, and expedient devices.	21	8	29	(24)
Squad Tests	Review and exercise covering all instruction.	23	16	39	(37)
Platoon Tests	Review and exercise covering all instruction.	42	24	66	(63)
	Total hours within master training program.	210	150	360	(314)

NOTES:

a. Maximum number of trained indigenous personnel will be used to assist in training others. Identify those personnel who may qualify as potential cadre or potential leaders.

b. Intelligence, compass, map familiarization, observing and reporting, tactical training of the individual, patrolling, weapons, demolitions, and field sanitation will be integrated whenever possible.

c. Classes to be broken down into platoon size groups when ever possible.

d. Practical work exercises, demonstrations, and conferences to be used in lieu of lectures to the maximum extent possible.

e. Stress small unit training (patrol, squad, and platoon). Develop teamwork and esprit-de-corps.

APPENDIX IX

ATOMIC DEMOLITIONS MUNITIONS SOP

(CLASSIFICATION)
              XXth SFG
              Fort Bragg, N.C.
              (date)

ANNEX—Demolition Munitions (ADM)

1. APPLICATION

This SOP supersedes all previous SOP's and applies except as modified by Group orders. Subordinate unit SOP's will conform, Attached units will comply with this SOP.

2. REFERENCES

(List SOP, directives, or policies of higher headquarters on which this SOP is based.)

3. ADMINISTRATION

  a. Responsibilities.

    (1) For preparation and periodic review of the SOP.

    (2) For requisitioning and posting of changes to all pertinent publications.

(CLASSIFICATION)

(CLASSIFICATION)

b. Records and Reports.

   (1) Target acquisition/damage assessment report.

   (2) Other required reports to higher headquarters.

   (3) List required test and maintenance records as specified by pertinent manuals.

   (4) Instructions on Post Strike Reports.

4. PUBLICATIONS

   a. Requisitioning Procedures.

   b. Reference Material, appendix I.

   c. Unsatisfactory Reports (UER and UR).

   (1) Preparation and study of proposals.

   (2) Submission and review procedures.

   (3) UR and UER File.

5. SUPPLY

   a. Allocation, Delivery, and Receipt.

   b. Authority (TOE, TA, TD, EMR).

   c. Equipment Lists (refer to the appropriate parts list).

   d. Requisition Procedures (including local requirements).

   e. Property Accountability (including local requirements).

(CLASSIFICATION)

(CLASSIFICATION)

6. SAFETY

   a. General.
      (1) A statement allowing no deviation from the approved checklist.
      (2) Applicable safety requirements deemed necessary.
   b. Electrical Safety Requirements.
   c. Explosive Safety Requirements.
   d. Nuclear Safety Requirements.
   e. Disposal of Contaminated Material.

7. TRAINING

   ADM training should be conducted in the following areas:

   a. Prefire Procedures.
   b. Site Preparation.
   c. Formal Instruction.
      (1) Classroom presentation.
      (2) Manual study.
      (3) UR and/or UER preparation.
      (4) Review of SOP and demolition fire orders (DFO).
   d. Support Training.
      (1) Emplacement site preparation.
      (2) Team organization.
      (3) Site security.

(CLASSIFICATION)

(CLASSIFICATION)

8. SECURITY

 a. Statement of Policy.

  (1) Importance.

  (2) Possible consequence of violations

  (3) Responsibilities.

 b. Security Measures to be Taken at Departure Airfield or Pickup Site.

 c. Security and Storage in Friendly Territory and in the GWOA.

 d. Receipting for ADM.

 e. Handling of Combinations.

 f. Document Control.

 g. Classified Item Control.

 h. Classified Study Procedures.

 i. Clearances—appendix 2.

 j. Access List—appendix 3.

9. TRANSPORTATION

 a. Air Movement. Procedures to acquire ADM, place aboard aircraft, and following actions required.

 b. Personnel and Duties. (List duties and required clearances.)

 c. Control.

  (1) Movement coordination.

(CLASSIFICATION)

(CLASSIFICATION)
- (2) Coordination with other forces.
- (3) Procedures in case of unavoidable delay or mechanical breakdown. (Other than an accident or incident.)

10. ORGANIZATION FOR ADMISSIONS

This paragraph outlines the capabilities of the deployed operational detachment, caching techniques, and suggested organization of the ADM firing party. Mission in support of allied forces will require modifications.

a. Team Leaders. (Indicated by position rather than name.)

    EXAMPLE: CO, EXO, OP Sgt, DML Sgt, etc.

b. Composition and Duties.
- (1) Prefire Team (for composition see table 1).
  - (a) Pickup of ADM equipment, appendix 4.
  - (b) Transportation procedures.
  - (c) Prefire procedures, appendix 5.
  - (d) Remote command fire procedures, appendix 6.
  - (e) Basic immediate security of munition.
  - (f) Emergency disarm procedures, appendix 7.
- (2) Support team (size dependent on type of mission).
  - (a) Pickup and transportation procedures (mines, camouflage, etc.).

(CLASSIFICATION)

(CLASSIFICATION)

- (b) Preparation of the emplacement site (construction, installation of mines; wire; booby traps; etc.).
- (c) Preparation of remote command site(s).

(3) Security team (size dependent on terrain, tactical situation, etc.)
- (a) Establish emplacement site security before the arrival of munitions.
- (b) Provide security at the completed emplacement site until prearranged departure time.
- (c) Provide security detail at the command site until after detonation.

11. ACCIDENT AND INCIDENT PLAN

This paragraph will cover such contingencies as accidents, incidents, or delays, to include explosions, nuclear contamination, misfire malfunction, and damage.

a. General. It is recommended that a set of code words be prepared, if not already accomplished by higher headquarters, to allow understanding of the situation over unclassified means of communication.

b. Accident.
  (1) Definition. (The definition of an accident may be found in the Munitions Prefire Manual.)

(CLASSIFICATION)

(CLASSIFICATION)

- (2) Immediate action. (List local and higher headquarters requirements in full detail when possible.)
- (3) Notification. (Person to be notified by name or duty and location.)
- (4) Continuing action. (Protective measures, security, or control.)
- (5) Followup and reports.

c. Incident.
- (1) Definition. (The definition of an incident may be found in the Munition Prefire Manual.)
- (2) Immediate action. (List local and higher headquarters requirements in full detail when possible.)
- (3) Notification. (Person to be notified by name or duty and location.)
- (4) Continuing action. (Protective measures, security, or control.)
- (5) Followup and reports.

12. EMERGENCY DISPOSAL AND DESTRUCTION

a. Priorities of Denial.

b. Authority for Emergency Disposal and Destruction.

c. Methods of Disposal.

CLASSIFICATION

CLASSIFICATION

d. Methods of Destruction.

e. List of Materials Needed.

APPENDIXES:

Appendix 1 List all FM's, TM's, TC's Ordnance Special Weapons Technical Instruction, etc., necessary for ADM firing party personnel to scan, read, or study, and the frequency of revision.

Appendix 2 List all cleared personnel with the Group indicating their proper clearance.

Appendix 3 Include the permanent access list and the procedures for escorting visitors into the emplacement site.

Appendix 4 Include here a checklist of pickup requirements. (Signature cards, SASP access requirements, documents, forms and records, ramps, lifting devices, tie down equipment, etc.).

Appendix 5 Include here a prefire checklist for each munition.

Appendix 6 Include here a checklist for remote command site. (Location of foxholes, security, firing procedures, change of mission procedures, etc.).

CLASSIFICATION

CLASSIFICATION

Appendix 7   Include here a checklist for the emergency disarming of all munitions for which there is a checklist in appendix 5.

CLASSIFICATION

GLOSSARY

This glossary is provided to enable the user to become familiar with certain terms unique to Special Warfare. Further reference may be made to JCS, part I, Dictionary of United States Military Terms for Joint Usage, dated December 1964, AR 320–5, and FM 31–21.

AST (area specialist team)—An area specialist officer and an area specialist NCO who assist in precommitment planning of detachments, coordination of detachment activities in the isolation area, and follow through on all messages to and from committed detachments.

BLIND DROP—Technique wherein selected U.S. and indigenous personnel are airdropped during the initial infiltration phase on drop zones devoid of reception personnel.

CARP (computed air release point)—System used when no ground personnel are available to provide terminal markings. The desired impact point must be known and landmarks, short of the impact point and in the line of flight from the initial point, must be visually identified. CARP is the parachute release system normally used during joint airborne operations.

CSS (catalog supply system) — Comprises packages of associated individual items as well as units comprising several packages, permitting users maximum flexibility in requesting supplies. Broad classification of supplies are made and code designations are assigned to each. The coding system is not secure but reduces message lengths when a variety of supplies are ordered.

GWOA (guerrilla warfare operational area)— Geographical area in which the organization, development, conduct, and supervision of guerrilla warfare and associated activities by Special Forces detachments assist in accomplishing the theater mission.

HALO (high altitude low opening)—Means of infiltration used when parachute entry from high altitude is preferred. Freefalling to a low altitude ensures greater accuracy in reaching the desired impact point and the delayed parachute opening minimizes detection by the enemy.

LOLEX (low level parachute opening extraction) —Method of delivering materiel on the DZ with the load being extracted from the aircraft at an altitude between 2 to 6 feet by opening a parachute or parachutes attached to the load.

RECEPTION COMMITTEE—Committee formed to control the DZ or landing area by providing security for the reception operation, emplacing DZ markings, maintaining surveillance of the DZ before and after the reception operation, recovering and dispersing incoming personnel and cargo, and sterilizing the site.

RCL (reception committee leader)—Leader of the reception committee, usually consisting of five parties, i.e., command party, marking party, security party, recovery party, and transport party.

SKYHOOK. Skyhook operations are the long-range, high-speed, aerial pickup of men and materiel from remote or inaccessible areas, by specially equipped aircraft traveling at a safe altitude and at a maneuverable speed.

STERILIZATION. As used in this manual, sterilization means elimination or obliteration of all traces or evidence that an area or site has been occupied or used for any purpose.

METRIC CONVERSION TABLE

Linear Conversion—English-Metric Systems

METRIC CONVERSION TABLES
Linear Conversion – English-Metric Systems

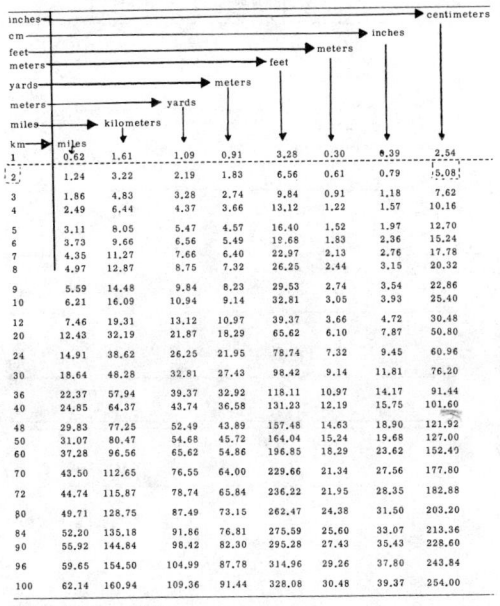

inches→km	miles→miles	kilometers	meters→yards	yards→meters	meters→feet	feet→meters	cm→inches	inches→centimeters
1	0.62	1.61	1.09	0.91	3.28	0.30	0.39	2.54
2	1.24	3.22	2.19	1.83	6.56	0.61	0.79	5.08
3	1.86	4.83	3.28	2.74	9.84	0.91	1.18	7.62
4	2.49	6.44	4.37	3.66	13.12	1.22	1.57	10.16
5	3.11	8.05	5.47	4.57	16.40	1.52	1.97	12.70
6	3.73	9.66	6.56	5.49	19.68	1.83	2.36	15.24
7	4.35	11.27	7.66	6.40	22.97	2.13	2.76	17.78
8	4.97	12.87	8.75	7.32	26.25	2.44	3.15	20.32
9	5.59	14.48	9.84	8.23	29.53	2.74	3.54	22.86
10	6.21	16.09	10.94	9.14	32.81	3.05	3.93	25.40
12	7.46	19.31	13.12	10.97	39.37	3.66	4.72	30.48
20	12.43	32.19	21.87	18.29	65.62	6.10	7.87	50.80
24	14.91	38.62	26.25	21.95	78.74	7.32	9.45	60.96
30	18.64	48.28	32.81	27.43	98.42	9.14	11.81	76.20
36	22.37	57.94	39.37	32.92	118.11	10.97	14.17	91.44
40	24.85	64.37	43.74	36.58	131.23	12.19	15.75	101.60
48	29.83	77.25	52.49	43.89	157.48	14.63	18.90	121.92
50	31.07	80.47	54.68	45.72	164.04	15.24	19.68	127.00
60	37.28	96.56	65.62	54.86	196.85	18.29	23.62	152.40
70	43.50	112.65	76.55	64.00	229.66	21.34	27.56	177.80
72	44.74	115.87	78.74	65.84	236.22	21.95	28.35	182.88
80	49.71	128.75	87.49	73.15	262.47	24.38	31.50	203.20
84	52.20	135.18	91.86	76.81	275.59	25.60	33.07	213.36
90	55.92	144.84	98.42	82.30	295.28	27.43	35.43	228.60
96	59.65	154.50	104.99	87.78	314.96	29.26	37.80	243.84
100	62.14	160.94	109.36	91.44	328.08	30.48	39.37	254.00

Example: 2 inches = 5.08 cm

AGO 6242C

Linear Conversion (continued)

One Unit (below) Equals ⟶	mm	cm	meters	km
mm (millimeters)	1.	0.1	0.001	0.000,001
cm (centimeters)	10.	1.	0.01	0.000,01
meters	1,000.	100.	1.	0.001
km (kilometers)	1,000,000.	100,000.	1,000.	1.

One Unit (below) Equals ⟶	gm	kg	metric ton
gm (gram)	1.	0.001	0.000,001
kg (kilograms)	1,000.	1.	0.001
metric ton	1,000,000.	1,000.	1.

Units of Centimeters

cm	1	2	3	4	5	6	7	8	9	10
Inch	0.04	0.08	0.12	0.16	0.20	0.24	0.28	0.31	0.35	0.39

Fractions of an Inch

Inch	1/16	1/8	3/16	1/4	5/16	3/8	7/16	1/2
cm	0.16	0.32	0.48	0.64	0.79	0.95	1.11	1.27

Inch	9/16	5/8	11/16	3/4	13/16	7/8	15/16	1
cm	1.43	1.59	1.75	1.91	2.06	2.22	2.38	2.54

Linear Conversion—Continued

(1) Weight Conversion - English-Metric System

	short ton	metric ton	pounds	kilograms	ounces	grams
metric ton(3) ←						
short ton(2) ←						
kg ←						
pounds ←						
grams ←						
ounces ←						
1	1.10	0.91	2.20	0.45	0.04	28.4
2	2.20	1.81	4.41	0.91	0.07	56.7
3	3.31	2.72	6.61	1.36	0.11	85.0
4	4.41	3.63	8.82	1.81	0.14	113.4
5	5.51	4.54	11.02	2.67	0.18	141.8
6	6.61	5.44	13.23	2.72	0.21	170.1
7	7.72	6.35	15.43	3.18	0.25	198.4
8	8.82	7.26	17.64	3.63	0.28	226.8
9	9.92	8.16	19.84	4.08	0.32	255.2
10	11.02	9.07	22.05	4.54	0.35	283.5
16	17.63	14.51	35.27	7.25	0.56	453.6
20	22.05	18.14	44.09	9.07	0.71	567.0
30	33.07	27.22	66.14	13.61	1.06	850.5
40	44.09	36.29	88.18	18.14	1.41	1134.0
50	55.12	45.36	110.23	22.68	1.76	1417.5
60	66.14	54.43	132.28	27.22	2.12	1701.0
70	77.16	63.50	154.32	31.75	2.47	1984.5
80	88.18	72.57	176.37	36.29	2.82	2268.0
90	99.21	81.65	198.42	40.82	3.17	2551.5
100	110.20	90.72	220.46	45.36	3.53	2835.0

Example: Convert 28 pounds to kg
    28 pounds = 20 pounds + 8 pounds
    From the tables: 20 pounds = 9.07 kg and 8 pounds = 3.63 kg
    Therefore, 28 pounds = 9.07 kg + 3.63 kg = 12.70 kg

(1) The weights used for the English system are avoirdupois (common) weights.
(2) The short ton is 2000 pounds.
(3) The metric ton is 1000 kg.

Weight Conversion—English-Metric Systems

Volume Conversion – English-Metric Systems

cu. ft	→ cu. yd	→ cu. meters	cu. ft	→ cu. meters	→ cu. ft	→ cu. yd
1	0.037	0.028	27.0	0.76	35.3	1.31
2	0.074	0.057	54.0	1.53	70.6	2.62
3	0.111	0.085	81.0	2.29	105.9	3.92
4	0.148	0.113	108.0	3.06	141.3	5.23
5	0.185	0.142	135.0	3.82	176.6	6.54
6	0.212	0.170	162.0	4.59	211.9	7.85
7	0.259	0.198	189.0	5.35	247.2	9.16
8	0.296	0.227	216.0	6.12	282.5	10.46
9	0.333	0.255	243.0	6.88	317.8	11.77
10	0.370	0.283	270.0	7.65	353.1	13.07
20	0.741	0.566	540.0	15.29	706.3	26.16
30	1.111	0.850	810.0	22.94	1059.4	39.24
40	1.481	1.133	1080.0	30.58	1412.6	52.32
50	1.852	1.416	1350.0	38.23	1765.7	65.40
60	2.222	1.700	1620.0	45.87	2118.9	78.48
70	2.592	1.982	1890.0	53.52	2472.0	91.56
80	2.962	2.265	2160.0	61.16	2825.2	104.63
90	3.333	2.548	2430.0	68.81	3178.3	117.71
100	3.703	2.832	2700.0	76.46	3531.4	130.79

Example: 3 cu. yd = 81.0 cu. ft

Volume: The cubic meter is the only common dimension used for measuring the volume of solids in the metric system.

Volume Conversion—English-Metric Systems

INDEX

	Paragraph	Page
Accompanying supplies	120a	261
Accountability, supplies	121	263
Acid, picric	144b(3)	321
Acid, sulfuric	144a(4)	321
Administrative plan (FTX)	App IV	389
Advanced charges	135	274
Aggressor forces	App IV	389
Air and amphibious messages	App III	384
Aircraft capabilities	27	37
Air delivery containers	27	37
Air landing operations	75–85	128
Air operations (Gen):		
Release point	63	107
Planning	27, 16, app IV	37, 23, 389
Characteristics	54a	83
Resupply	55	85
Alternate drop zone	58	102
Ambushes:		
Explosives	136d	291
Small boat	107c	217
Special	90c	187
Water	90c(3)	190
Ammonia dynamite	136f, 144b	294, 318
Ammonium nitrate	136f	294
Amphibious operations	92, 93	194, 195
Animals, domestic	App V	449
Antennas	112	230
Area, GWOA	4, 33–41	6, 52
Area Study/Assessment:		
Accompanying supplies	120a	261

	Paragraph	Page
Area Study/Assessment—Continued		
Air operations in counterinsurgency.	31	46
Area (general)	6b	10
Automatic resupply	120b	621
General	6a	9
GWOA intelligence	6c	10
Operational use of intelligence in unconventional warfare.	7	10
PSYOPS	15b(1)	21
Arthropods	166e, app V	354, 449
Assault element	89b(2)	168
Assembly, ground	27a(5)	40
Assembly point, alternate	27a(5)	40
Atomic Demolitions Munitions (ADM).	153, 154; app IX	338, 503
Audience, PSYOPS (target)	19	26
Audible Signal Radio	110d	225
Automatic resupply	120b	261
Bangalore torpedo	136d	291
Base, GWOA	37b	61
Beach landing sites	97a	201
Binoculars	27a(3)	38
Black powder explosive	139e	305
Blasting caps	135b, 145b	276, 324
Blind drop	26, 27c	37, 42
Boat, inflatable	92b	194
Boat operations, small:		
Advantages	105a	208
Counterinsurgency	104	207
Landing requirements	105b(4)	210
Limitations	105b	209
Medical evacuation	107e(1)	220
Patrol	105	208
Reconnaissance patrols	107b	214
Tactics	107	213

	Paragraph	Page
Booby traps	150b(2)	330
Border surveillance	8c, 48	14, 73
Breaching:		
Barbed wire	137f(1)	302
Mine fields	137f(2)	302
Radius	152c(1)	336
Brick, incendiary	141b	306
Brief back	7d	13
Briefing center	7c	12
Briefing preflight	27a(6)	41
Burning fuse	145c	324
Caches	30	46
Considerations	125b	266
Medical	164c	351
Mission support site (MSS)	126	266
Preemergency	125	266
Underwater	101c(2), (3)	206
Calculation, metric	152	333
Can, water delay	142f	311
Candle delay	142c	308
Capsule, gelatin	142e(4)	311
CARP	69	119
Catalog supply systems	App VII	482
Cavity	136a(2)	286
Celluloid crystal	142g(1)	313
Chemical formulas	138c	303
Chlorate, potassium	139	304
Cigarette delay	142b	308
Civic action	8c, 44b, 149	14, 68, 327
Civil defense groups	45c	70
Civil guard	45a	69
Civilian (GWOA)	38, 123	61, 264
Civilian supply support	123	264
Claymore, improvised	136e	291
Climatology	4, app VI	6, 465
Closed circuit SCUBA	94d	197
Coal dust	136c	289

	Paragraph	Page
Code:		
Crypto	114	250
Radio	110, 111	223, 228
Committee, reception	72, 73	121, 122
Communications:		
Animals	110g(2)	228
Antenna	112	230
Audible signal	110d	225
Code	111b	228
Counterinsurgency	115	253
Crypto	114d	252
Frequencies	111d	229
FTX	46, app IV	70, 389
GWOA	109b, c	222, 223
Local	110f	227
Maintenance	111c	229
Message	113	243
PSYOPS	25b(3), (5)	35
Radio	110b	224
Security	114, 118	250, 258
Small boat operations	106c	212
SOI/SSI	55b, 111d	90, 229
Telephone	110c	225
Training	111	228
Types	109	222
Unconventional warfare	109	222
Visual signal	110e	225
Wire	116	257
Compass	27a(3)	38
Cone	136a(2)	286
Construction	130	270
Containers, air delivery	27a(3)(a)	39
Control group (FTX)	App IV	389
Control plan (FTX)	App IV	389
Counterinsurgency	43–52	67
Communications	25a	34
Demolitions	145b	324
Intelligence	5, 8	8, 13
PSYOPS	25a	34

	Paragraph	Page
Counterinsurgency operations	31	46
Critique plan (FTX)	App IV	389
Cryptograph	114	250
Crystal, celluloid	142g(1)	313
CSS	54c, 120c, app VI	85, 262, 465
Debarkation	93, 94	195, 196
Demobilization (medical)	163, 171	349, 359
Demolitions:		
ADM	App IX	503
Ambush	136d	291
Calculations, metric	152	333
Control, resources	151	331
Counter force	135c	277
Counterinsurgency	147	325
Delay, explosive	142	308
Foreign	144	318
Handling	146	324
Improvised	136	284
Planning	133–154, app IX	272, 503
Saddle charge	135a	274
TNT	144b(1)	321
Train derailment	143	316
Demolition card (GTA 5-10-9)	152	333
Denial operations (Border)	8c, 48	14, 73
Derailment, train	143	316
Destructive techniques	134	273
Detonating cord	135, 143	274, 316
Diamond charge	135b	276
Domestic animals	App V	449
Drift, wind	63b(2)	107
Drop zone (DZ):		
Alternate	58	102
Area	60	103
Authentication	74	127
Description	56a	87

	Paragraph	Page
Drop zone (DZ)—Continued		
HALO	27b, 68a	42, 117
Marking	55b, 61, 62, 64–66, 68	86, 105, 107, 111, 119
Reporting	57	95
Responsibility	53	82
Safety factor	56b(1)(b)	90
Security	56b(3)	94
Selection	53a, 56	82, 87
Sterilization	73e(4)	125
Unmarked	66	115
Weather	56a, b	87, 90
Dust initiator	136c	289
Dynamite	144a	318
Electric firing (Demo)	137a(2)	296
Emergency plans	27a(6)	41
Emergency resupply	120d	262
Engineering expedient	148	326
Epidemiology	App VI	465
Equipment	27	37
Escape and evasion	App IV	389
Evacuation	132, 160a, app IV	271, 345, 389
Exercise area	App IV, 4	389, 6
Exercise control group	App IV	389
Exercise field	App IV	389
External logistics	119–121	260
Factor, material	152c	335
Fertilizer	136f	294
Field exercise (FTX)	App IV	389
Flour	139c	304
Foreign explosives	144	318
Formulas, chemical	138c	303
Fougasse	136d(2)	291
Free fall	27b	42
Fuse string	142d	308

	Paragraph	Page
Gasoline	138d	303
Gelatin gas	140a	305
Geography	App V	449
Grenades	136b	288
Ground assembly	27a(5)	40
Ground release point	63	107
Guerrilla warfare	32–41	52
Guncotton	144b(4)	321
GWOA	4	6
HALO	26, 31, 67	37, 46, 117
Handling foreign explosives	144–146	318
Helicopters	31	46
Homing device	62	107
Hospitalization	160, 132, App IV	345, 271, 389
Hydrography	97a	201
Igniters, incendiary mixture	142	308
Improvised claymore	136e	291
Incendiaries:		
Aluminum powder	139d, e	305
Brick	141b	306
Improvised	138	303
Thermite	141a	306
Indigenous forces:		
Assessment	App VI	465
Demolition training	134	273
Operations training	52	81
Training	49	79
Individual loads	27a(3)(c), 120c, app VII	39, 262, 482
Infiltration:		
Air	27, 31	37, 46
Air delivery containers	27a(3)(a)	39
Blind drop	27c	42
General	26	37
HALO	27b	42
Land	29, 31c	45, 50

	Paragraph	Page
Infiltration—Continued		
Reception committee	27a(2)	38
Ship-to-shore movement	28a(4)	44
Stay-behind	30, 31c	46, 50
Water	28, 31b	44, 49
Infrared	95	198
IP (initial point)	56a(1)(d)	88
Initial area assessment	6–8, app V, app VI	9, 449, 465
Initiator, dust	136c	289
Inner security element	73d	123
Insulators, wire	112a	230
Intelligence:		
Area assessment	9	15
Ambushes	90a(3)	183
Boat operations	106a	211
Equipment and supplies	13	17
FTX	App IV	465
Gathering	48f(7)	78
GWOA	4, 33, 41	6, 46, 63
Intelligence in counterinsurgency.	5	8
Medical	172–174	359
Operational use in counterinsurgency.	8	13
Operational area intelligence.	6	9
Photography	11, 12	16, 17
Psychological operations	10, 18	15, 25
Raids	89c(2)	171
Unconventional warfare	7	10
Interdiction	41, 91	63, 190
Internal logistics	122–129	263
Labor	131	271
Landing craft	92	194
Land infiltration	29	45

	Paragraph	Page
Landing operations:		
Debarkation techniques	94	196
Sites	96	200
Landing sites	97a	201
Alternate	97d	202
Marking	98a–e	203
Primary	97a, b	201, 202
Landing zone (LZ):		
Helicopter	81–83	149
Marking	78–83	136
Reporting	84	157
Responsibility	53	82
Security	76	128
Selection	77b	129
Water	80	140
Launcher rocker	137a	295
Leadership school	51	80
Lockout	94c	197
Logistics (external):		
Accompanying supplies	120(a)	261
Accountability (final)	121	263
Automatic resupply	120(b)	261
Catalog supply system	App VI, 120(c)	465, 262
Delivery	120	260
General	119	260
On call supplies	120c	262
Logistics (internal):		
Bulk food and clothing	122	263
Civilian support	123	264
Consumption factors	122b	264
Decentralized system	127	267
Geography consideration	122	263
Local procurement	124	265
Replacement factor	122a	264
Security	128c	269
Supply maintenance	129	269
Transportation	128	268

	Paragraph	Page
Logistical support:		
Counterinsurgency	48d(5)	76
FTX	App IV	389
Loudspeaker	25a, b	34
Low level parachute extraction resupply (LOLEX).	88	165
LZ	31, 75–85	46, 128
MAAG	8c	14
Magnesium	136c	289
Maintenance	129	269
Marking LZ	78	136
Marking DZ:		
Devices	27	37
HALO	27b, 68	42, 117
Methods	64	111
Placement	65	112
Security (fig. 16)	65	112
Unmarked	66	115
Master training program	App VIII	495
Media psychological operations	21, 22	29, 31
Medical:		
Capabilities	156b(1)–(3)	341, 342
Counterinsurgency	165, 166	351, 352
Epidemiology	App VI	465
Organization	156	341
Preventive techniques	157	342
Requirements	155	340
Supply	164, app VII	350, 482
Medical kit	27a(3)	38
Message, sample	App III	384
Message writing	113	243
Metric calculations	152	333
Military civic action. (*See* Civic action.)		
Mining	91	190
Missions, Counterinsurgency	44	67
Mission support site (MSS)	126	266

	Paragraph	Page
Navigation, amphibious	95	198
Nitrate, ammonium fertilizer	136f	294
Nitroglycerin	144b(5)	322
Non-electric firing	137c	297
Nuclear, Atomic demolition munition.	1, 153, app IX	5, 338, 503
Objective area	31a	47
On-call supplies	120c	262
Open circuit SCUBA	94c, 102, 103	197, 206, 207
Operational detachment	33, 54	52, 83
Operations, border denial	8c, 48	14, 73
Operations plan (FTX)	App IV	389
Opposed charge	135c	277
Outer security element	73d	123
Parachute	27	37
Patrols, small boat	105	208
People, area study	App V	449
Photography, equipment	27a(3)	38
Photography, intelligence and supplies.	11–13	16
Picric acid	144b(3)	321
Pigeon	110g	227
Placement of marking	65	112
Plan, administrative (FTX)	App IV	389
Plan, control (FTX)	App IV	389
Plan, critique (FTX)	App IV	389
Plan, intelligence (FTX)	App IV	389
Planning, operational (counterinsurgency).	31	46
Plastic explosives	135, 141b	274, 306
Platter charge	135d	281
Potassium chlorate	139	304
Potassium nitrate sulphur	139c	304
Potassium permanganate-aluminum.	139d	305
Powder-aluminum	139e	305

	Paragraph	Page
Pre-emergency caches	125	266
Preplanned air supply operations.	55	85
Preventive medicine	157, 158	342, 343
Primers, foreign	145a	322
Priming	135c(3)	281
Procurement, supply	124	265
Propaganda	20, 21, 24, 25	28, 29, 33
Propaganda, media	25	33
Psychological operations	14–25	19
Radios:		
Propagation	111d	229
Psychological operations	25a(3)	34
Raids:		
Assault element	89b(2)	168
Large	89g	177
Movement	89d	173
Security element	89b(3)	169
Small boat	107d	218
Withdrawal	89f	174
Rappelling, helicopters	31a(2)	48
Reception committee:		
Air	27a(2)	38
Functions and duties	72, 73	121, 122
Water	28a(2), 99	44, 204
Reconnaissance, boats	92b	194
Reconnaissance patrols, small boat.	107b	214
Recovery party	73a(4)	122
References	App I	363
Reference point, DZ	57a(6)	96
Rehearsals, infiltration	27a(4)	40
Release point	63–65	107
Reporting:		
DZ	57	95
Landing sites	97	201
LZ	84	157

	Paragraph	Page
Responsibility local supply procurement.	124	265
Resupply	31a	47
Automatic	120b	261
Emergency	120d	262
HALO	71	120
LOLEX	88	165
On call	120c	262
Ribbon charge	135e	283
Rockets	137a	295
Saddle charge	135a	274
Sawdust and wax	140a	305
Security:		
Communication	114, 118	250, 258
Crypto	114d	252
DZ	58	102
GWOA	39	62
LZ	76	128
Security element	89e	173
Self contained underwater breathing apparatus (SCUBA).	100–103	205
Self defense units	45b	69
Sewage disposal	App VI	465
SFOB	54	83
Shaped charge	136a	285
Ship-to-shore movement	28a(4)	44
Shoelace	142d	308
Skyhook	86	158
Sniping and mining	91	190
SOI/SSI	55	85
SOP (Adm)	App IV	389
Stand-off distance	136a(2)	286
Stay-behind operations	30	46
Steel, structural	135e, 152a	283, 333
Sterilization (DZ)	73e(4), (5)	125, 126
String fuse	142d	308

	Paragraph	Page
Strike forces	31a(1)	47
Submarine	92a, 94	194, 196
Sugar-potassium:		
Chlorate	139a	304
Permanganate	139b	304
Supplies:		
Accompanying	120a	261
Air	27a(3), 54	38, 83
Automatic	55a	85
Emergency	55b	86
Land	29	45
Medical	164	350
On-call	120c	262
Photography	13	17
Signal	6, app VII	9, 482
Small boat operations	104–107	207
Stay-behind	30	46
Transportation	128	268
Water infiltration	28	44
Supply drop zone	56	87
Surf zone	95c	199
Surface craft	94d	198
Tamping factor	152c(3)	338
Targets, adjacent to water	101b	205
Target audiences, PSYOPS	18	25
Telephone	109c, 110c	223, 225
Thermite	136c, 140a	289, 305
Thermite grenade	141a	306
Timber	152b	333
TOE equipment	27a(3)	38
Topography	App VI	465
TNT	136c	289
Track, DZ	59b	103
Train derailment	143	316
Training program	App VIII	495
Transportation	128, app IV	268, 389
Transporting vessel	92b	194
Trip wires	136d	291

	Paragraph	Page
Underwater operations	101	205
Unmarked DZ	66	115
Unconventional warfare operations.	27–31, 45	37, 68
Veterinary medicine techniques	167–174	355
Visual signals	110e	225
Watch delay	142g	313
Water operations:		
Debarkation	94	196
Equipment	102	206
Infiltration	28	44
Landing operations	95	198
Limitations	103	207
Planning	92	194
Reception committee	99	204
Reporting	97	201
Site marking	98	203
Site selection	96	200
Small boat (counterinsurgency)	104–107	207
Supplies	28a(3)	44
Tactics	90c(3), 93	183, 195
Underwater	101	205
Wax/sawdust	140b	306
Weapons	App VII	482
Weather, DZ	56a(2)	89
Wind drift	63b(2)	108